KU-252-832

Age
Proof

Professor Rose Anne Kenny is an award-winning physician and researcher who has been Head of the academic department of Medical Gerontology at Trinity College, Dublin since 2006. She is the founding Principal Investigator of The Irish LongituDinal study on Ageing (TILDA). Prof Kenny has published over 600 scientific publications to date and was admitted as a Member of the Royal Irish Academy (M.R.I.A) in 2014 – Ireland's highest recognition for scientific excellence. She recently received a Lifetime Achievement Award for her research on ageing at the world congress in Kuala Lumpur 2019. She was voted a Health Hero by the *Irish Times* in 2018, won the Trinity Innovation Award in 2017 and was elected President of the Irish Geriatrics Society in 2020.

Age Proof

The New Science of Living a Longer and Healthier Life

PROFESSOR ROSE ANNE KENNY

LAGOM
BOOKS FOR A BETTER BALANCED LIFE

First published in the UK by Lagom
An imprint of Bonnier Books UK
4th Floor, Victoria House, Bloomsbury Square,
London, WC1B 4DA
England

Owned by Bonnier Books
Sveavägen 56, Stockholm, Sweden

Hardback ISBN: 978 1 7887 0504 2
Trade Paperback ISBN: 978 1 7887 0505 9
Ebook ISBN: 978 1 7887 0507 3

All rights reserved. No part of the publication may be reproduced, stored in a retrieval system, transmitted or circulated in any form or by any means, electronic, mechanical, photocopying, recording or otherwise, without prior permission in writing of the publisher.

A CIP catalogue of this book is available from the British Library.

Designed and typeset by Envy Designd

Printed and bound by Clays Ltd, Elcograf S.p.A

1 3 5 7 9 10 8 6 4 2

Copyright © Rose Anne Kenny 2022

Rose Anne Kenny has asserted her moral right to be identified as the author of this work in accordance with the Copyright, Designs and Patents Act 1988.

Every reasonable effort has been made to trace copyright holders of material reproduced in this book, but if any have been inadvertently overlooked the publishers would be glad to hear from them.

Lagom is an imprint of Bonnier Books UK
www.bonnierbooks.co.uk

Dedicated to the memory of my Mum and Dad –
Kay and Billy Kenny

Contents

Preface

ON A WET, DARK NIGHT IN JANUARY 2018, I PLOUGHED through the pools of water that covered the black road heading for a town in the middle of Ireland. I was to give a lecture on ageing and health to the townsfolk. During the miserable journey, my despondency deepened at the hopelessness of ever attracting an audience on such a bad night. Hosted in a cold hotel which usually catered for funerals and weddings, it was billed as 'the first lecture of an all-Ireland tour to share new research by a Trinity College academic'.

The ballroom was large, cold and empty; the small lectern seemed out of place and lonely, staring down on a mass of empty gold wedding chairs. The overhead projector was too old to be compatible with our PowerPoint system, so my PA headed off into the night to source alternative technology. Mumbling to myself, 'I must be mad,' I encountered the shy hotel manager who completed my gloomy apprehension by apologising on account of a competing event across the street, the 'Mission'. I hadn't heard that term in so long. The Mission is a long-standing Irish tradition, an annual event where the local Catholic church hosts preachers and sermons from visiting religious orders. My heart sank: in rural Ireland, the Mission would be inequitable competition.

Yet little by little, the hall started to fill. People of all ages bustled in: 30-year-old mums with kids, 50-, 60-, 70-year-old men and women. Two buses pulled up outside and disgorged their prattling contents from the surrounding countryside and villages. Next, residents from a care home shuffled in, transported by a kindly local volunteer police officer in full uniform. The space began to warm with mounting chatter, growing laughter and the clinking of china. The local GAA football club was serving tea, coffee and cakes, and on display were Ireland's two most prized sporting trophies: the Sam Maguire and Liam McCarthy cups, provoking photos and more banter from the swelling crowd. A local children's band set up their instruments; the audience took their seats and, to the rhythm of lively jigs and reels, I settled in to begin my first of many such lectures.

Afterwards, I met with the audience, who were brimming with questions and comments. I was taken aback when some remarked that they had never attended a lecture previously, 'apart from the priest's sermon on a Sunday morning' (ironic, in light of the Mission across the road they had now missed). This was chastening. Many asked if I had written down the content of the lecture and whether there was a book that covered the information I had shared, thus planting the seed for this publication. The book is a distillation of the content of the lectures and a tribute to the joy I experienced in sharing the knowledge and experience on the road trip of a lifetime.

◆

Time and time again, patients, work colleagues and friends have told me how they hate the thought of growing old. People in their forties and fifties say they try not to think about it because for them it carries such negative overtones. And yet, the science in this field is vast and moving ahead very rapidly. When I was a young doctor, it was almost non-existent but over the past 20 years it has exploded. The field continues to evolve at a pace, providing us with good evidence that the 'last lap', as one of my patients calls it, can actually be the most relaxed, worthwhile and enjoyable period of our lives – particularly if we prepare for it.

Part of that preparation is understanding what factors determine ageing and what we can do about it, in a timely fashion. Have you ever paused to wonder why we are living longer? A baby girl born today will live on average three months longer than her sister born last year. In 1800, you could expect to live to 40; 200 years later, that has more than doubled and we can expect to live to 85 and beyond. When I was starting my medical career, it was an unusual event to have a patient of 100 years or more in the hospital and we would flock to view this rarity – this is not uncommon today.

I was first wooed by clinical ageing when I was a young trainee doctor and the fascination for why we age continues to drive my curiosity and research. Then and now, engaging with patients and learning from their life stories guides answers and solutions, raising the obvious questions about why some people appear resilient to ageing while others 'age early'.

The Blue Zones hold many of the secrets that can help answer these questions. They are five places around the world, in Sardinia, Italy; Okinawa, Japan; California, USA; Nicoya, Costa Rica and Ikaria, Greece, all beside the sea, which have the highest proportions of centenarians worldwide. People of the Blue Zones don't just live longer but are fitter and stronger and less likely to get sick in old age. They are more likely to live healthy and physically active lives beyond 100.

In this book, I have built on the knowledge we've gained

from studying the Blue Zones to share the up-to-date science underpinning successful ageing. The cornerstone of successful longevity in the Blue Zones comprises several fun things: having a purpose in life, being curious, having lots of variety, laughter, friendship and enjoying a sense of belonging and close, strong connections with friends and family, including shared meals, wine and much more. Since the discovery of the zones and the factors that determine the healthy longevity of those who live there, there has been much research into why these things affect ageing and the biological reasons that underpin why Blue Zone people live so long and so well. How could something like having a purpose each day affect us biologically so that cell ageing slows down? And why have we evolved to need to have a purpose to survive? Knowing this, how can we ensure that we retain purpose in our lives every day? These are some of the questions that will be explored in these pages.

My book takes a whistle-stop tour of the key up-to-date learnings from my cumulative experiences as a clinician and as a researcher in this field. What makes this book unique is the distillation of cutting-edge research taken from one of the most comprehensive multidimensional research studies worldwide (which I lead), coupled with over 35 years of clinical experience in ageing medicine, illustrated by colourful patient histories collected over the years.

I have been privileged to establish and direct a groundbreaking research study of ageing, which has followed almost 9,000 adults aged 50 and older. Since 2009, the Irish Longitudinal Study on Ageing – TILDA – has generated over 400 research publications. The study covers all aspects of life, from sexual activity to food, to physical health and brain health, genetics, childhood experiences, expectations, friendships, finances and much more, to inform the complex picture that explains why and how we age. No one aspect drives ageing; it is a combination of factors – and many are within our reach to manipulate.

Drawing on TILDA and a host of other similar sister studies, I have been careful to ensure that the information included here is rigorously evidence based – no 'fake news'. I am as clear as I can be of the strength of evidence behind the information and shy away from anything that is conjecture. I stress this because a good friend in the States recently directed me to a 'brilliant' book (a bestseller) on a health and well-being topic. She did this in good faith, thinking that I would get 'inspiration' from the writing. I began reading it but I never finished the book as many of the definitive statements made by the author were based on supposition and not evidence based. I was taken aback at how my well-read friend could have been so gullible.

There is another reason I wanted to write this book. During my years as a clinician and researcher, I have witnessed a welcome transition in patients' expectations and curiosities. People are much better informed and consequently more engaged in the process of diagnosis and treatment, with both patients and practitioners moving slowly but surely towards shared decision-making and deeper awareness of a holistic approach to health and ageing. The medical profession more and more incorporates 'quality of life' and determinants of overall well-being into discussions with patients. The profession has gradually stepped out of traditional clinical silos and learned to interrogate wider life experiences that contribute to diseases and age-related processes. In my early days as a physician, medicine was much more didactic: the doctor 'told you what to do'. Part of the change in culture has come from sharing wider knowledge of all of the components that make up successful treatment responses, including lifestyle, relationships and attitudes.

A stark memory from my early days in training abides. It was the morning of a traditional mammoth ward round, with a senior clinician, a nurse, three junior doctors and two medical students all circled around a patient's bed on an open 16-bed ward – a daunting sight for any patient. The consultant stood at the head of the bed, his back to the stroke patient in question, holding court and waxing lyrical while he explained that 'this patient' was paralysed on the left side (arm and leg); that she was unlikely to ever recover and that her mental abilities were also affected because of the extent of the damage, all of which he had interpreted from her brain scan. He continued that she would not be able to live independently and would most probably have to go to a nursing home. The patient then sat up, admonished him for speaking over her, saying: 'I am right here under your nose, listening to all you say, so please address your statements to me. Yesterday, I started to move my left hand and arm and I took four steps with the help of a nurse. I have a very large and supportive family and I plan to go home, where they have already started to make changes to my house. I am a successful artist and I *will* paint again.'

I want to cheer when I recall her spirit and energy; it still gives me a rush of joy. Today, involving people in every step of care is commonplace and this is assisted by the availability of information on the internet. Practitioners are much better trained in communication. When presenting choices and offering full information, we gain a deeper understanding of the person, of what matters to them, why it matters, their expectations and the life experiences that shaped their presentation, decisions and thus our joint approach. More and more patients want to know why an illness or disorder occurs, to understand the background science of what may have gone wrong with their biology and to use this information to shape their decisions. So here, I have coupled how clinical disorders present as we age with the background biology that drives these changes.

I never ask a patient their age but rather base decision-making on an assessment of a person's biological age derived from a traditional physical examination and history. No two 83-year-olds are the same: one can run a marathon and the other is a frail nursing home resident. Making a clinical treatment judgement is very different for each and not related merely to a number. Our experiences and circumstances in childhood years all contribute to biology of middle age and later years.

In fact, biological ageing starts very early on – by our thirties, it has become established in cells. By reading this book, you will discover the breadth of biological ageing and how much it varies from chronological age. Biological age can be measured by internal 'biological clocks'. One study demonstrated a difference of 20 years in biological ageing clocks in adults as young as 38. Thus, age is not a number: our biological changes are what counts and the good news is that most of the factors that change our clocks are within our control to modify and improve – we control 80 per cent of our ageing biology. I have included some of the tests that we used in TILDA coupled with the expected normal results for your age and sex at the end of the book so that you can test yourself and see how you perform in measures which we know influence the pace of ageing.

This book explores and details the centuries-long search by mankind for the elixir of youth and life. I am excited to share good scientific evidence to persuade you that you *are* as young as you feel, that there is a multitude of things that can be done to enhance enjoyment of the 'last lap' and ensure lifelong satisfaction, curiosity and pleasure.

Chapter 1

You Are as Young as You Feel – Age is Not a Number

F OR ALL OF MY PROFESSIONAL LIFE I HAVE BEEN FASCINATED
by how people's attitudes influence not just ageing but their
health. Recently, I had an 85-year-old patient with a mild chest
infection who was very keen to return to full health speedily
because she provides daily assistance to, in her words, her 'elderly
neighbour'. It transpired that her neighbour was 74 but frail and
dependent on my patient who was more than happy to assist
with care. It amused me to hear the patient describe someone
11 years her junior as elderly, while clearly not perceiving herself
as such. She is typical of many people who do not 'feel their
age', who believe themselves to be younger than their years, than
their 'chronological number'. For them, the popular mantra that
today's 70-year-old is yesterday's 60-year-old is a truism, and
this attitude aligns well with the current science.

Eileen Ash is another great example of this. At the time of
writing, she is one of the oldest women in Britain and still driving
at age 105, having passed her driving test 80 years previously.

When I read about Eileen, I was struck by her positive attitude and that she had always led and continued to lead an active and varied life. Despite being over a century old, Eileen continues to take daily brisk walks and does yoga – an activity that she took up in her nineties, an age when most choose to slow down. She says, 'Some days I fancy doing the cat. Other days I do the cat and the dog. It makes my body feel much better. It keeps my muscles going.' She demonstrates a positive, optimistic attitude with plenty of courage and self-belief, which has empowered her to take up new challenges at each stage of life without being inhibited by age. She doesn't 'act her age' but rather continues to enthusiastically enjoy a full life; her chronological age does not interfere with her ambitions and approach to life and living.

Eileen is a living embodiment of how one's attitude towards age affects the speed at which biological ageing occurs. The science shows that her attitude helps to slow down both physical and cognitive ageing. My research group has done some interesting work in this area and shown how young or old one feels actually influences how quickly we age. In other words, the cellular processes which characterise ageing can be controlled by attitude and perceptions.

From the yoga-loving 105-year-old to the 40-year-old who struggles to run a mile, we all know people who seem surprisingly young or old for their age. We can distinguish between two forms of age that helps to explain this discrepancy: chronological age is measured from birth to a given date. Biological age, also called physiological age, is a measure of how well or poorly the body is functioning relative to chronological age.

We are born with a fixed number of genes – our DNA – but some of our genes can be switched on or off by factors such as diet, exercise and psychological approach and attitude. This switching on or off is called epigenetics. Biological ageing is defined by epigenetics, which occurs at all ages, and these changes in gene function speed up or slow down cell ageing. It explains the differences between biological and chronological ageing and why Eileen, at 105, appears and behaves 'younger' than others who are chronologically younger. Thanks to her positive attitude and because she has not stopped exercising throughout her life, Eileen has switched on protective genes which slow the pace of cell ageing. Epigenetics also explains why genetically identical twins who have the same genes but different life experiences and health behaviours age at different rates. Cells are more vulnerable to or better protected from damage depending on which genes are switched on or off.

We can measure epigenetics from blood samples and use the results to better understand why some people, like Eileen, live longer healthy lives. For example, our research has shown that adverse childhood experiences, such as parental alcoholism or household poverty, and mental health problems, such as depression, poor diet and low education attainment, show up in our genes and link to health problems in later adult life. Measuring epigenetics means we can see how changes to our genes are driven by modifiable life factors – factors that both we as individuals and society can influence to control our biological ageing and therefore how long we live for. In other words, epigenetics explains the link between an individual's attitude to age and actual cell ageing. To dig deeper into the science behind this and unlock some of the secrets behind successful ageing, we will first examine one of the world's most important recent scientific achievements – the human genome.

◆

In June 2020, we celebrated 20 years since the Human Genome Project was launched. Thanks to the work done on this project, we can better understand the genetic changes which determine Eileen's longevity. The Human Genome project was described by Tony Blair at its launch as 'a revolution in medical science whose implications far surpass even the discovery of antibiotics'. Then-US President Bill Clinton rather more dramatically stated, 'Today we are learning the language in which God created life.' It was a revolutionary scientific endeavour of enormous scope and scale.

Every one of us carries about 2 metres of DNA in every single cell, and we have 30 trillion cells. DNA consists of 23 pairs of chromosomes, each made up of 3 billion 'letters' of genetic information. The Human Genome Project was set up to read all of these letters. There is no index, no annotation and no easily discernible manner in which to navigate this obscure alphabet. It took thousands of scientists all over the world working together and sharing information every time they did a study over seven years to reveal more of the letters of the alphabet. It was slow, laborious and complicated. But, after 4 billion years of evolution, one organism – us – has been able to work out its own code of instructions. This has helped greatly with not just the diagnosis of genetic disorders but also in understanding the genes that contribute to longevity. Furthermore, we now understand a lot about the switching on and off of genes and how epigenetics is controlled by health behaviours and other external factors.

To date, one of the most prominent genes discovered to influence the ageing process is the DAF2 gene. Activity of this gene – i.e., whether it is switched on or off – controls many of the important pathways which govern how cells age. Examples of the role of this gene are also evident in animals. Manipulating the gene in animals – something which is not yet appropriate in humans – allows us to study how small changes in the gene function, and thus epigenetics, affect cell ageing and lifespan.

In species such as the worm, a small change in the DAF2 gene doubles lifespan. Because we share a large number of genes with the worm, this is likely to be the same for humans. DAF2 also controls insulin and growth hormone activity, both of which play key roles in the growth of all tissues and how we metabolise sugar and produce energy, both fundamental processes for survival of all cells. What's more, people who live to 90 and over have different DAF2 genetics than those who do not live beyond 90. Diet, obesity and exercise and caloric restriction influence the DAF2 gene, which may explain why these factors slow the pace of ageing and extend lifespan. This is an open door for us to use this new information in order to have more control over ageing.

. Epigenetic clocks emerged from the work of the Human Genome project and are an extension of what we know about epigenetics. When we refer to a gene being switched off or on, we are describing 'DNA methylation' – this is the addition of a methyl group to DNA (a methyl group is one carbon atom bonded to three hydrogen atoms). This occurs all the time throughout the body and helps to keep DNA stable. The amount of change in methylation can be used to determine tissue age. By charting this change throughout life, we have created the epigenetic clock as a measure of biological ageing. It is still a developing science and new 'clocks', which use different combinations of measures of methylation, continue to be discovered and tested for precision. No clock is as yet precise enough to clearly measure an individual's biological age but we are getting closer to this degree of accuracy. Very soon, we will be able to determine an individual's exact biological age.

So, in essence, the epigenetic clock allows for calculation of the difference between chronological and biological age – the pace of ageing. There has been recent hype surrounding this and there are now products on the market that claim to accurately determine biological age. At the time of writing, in my view, they should be approached with caution. Our research shows that, as yet, the methods are not sufficiently sensitive or specific to give accurate estimates of an individual's biological age and do not take into consideration all of the complex network of factors that influence the ageing process. But this is a fast-evolving area of research and no doubt more accurate tests of biological age are quickly coming down the track.

In recent years, we have learned much more about the many factors that affect epigenetic clocks. Those that adversely affect our clocks are illness, bad health behaviours (smoking or obesity) and stressful life experiences. Age acceleration occurs when our clocks speed up as a consequence of these events or behaviours. Another area that influences biological ageing is mood. The Canadian singer-songwriter Justin Bieber sleeps in a hyperbaric oxygen chamber, allegedly to relieve anxiety. Perhaps this is not as outlandish as it first appears. Persistent stress and mood change, such as depression and anxiety, can cause long-term damage from over-exposure to stress hormones and the adverse physiological state they create. A well-known New Zealand study, the Dunedin Study, followed 1,000 participants, all born between April 1972 and March 1973, with detailed testing at regular intervals since birth. At ages 26, 32 and 38, detailed health checks were carried out and blood tests were measured for biological ageing. This was coupled with details of the participants' perceptions of how they were ageing – their ageing attitudes. David Belsky and Terrie Moffitt, the lead investigators for the study, reported that some of the 38-year-olds had the epigenetic biological age of a 28-year-old, whereas others had the biological age of a 48-year-old (see opposite).

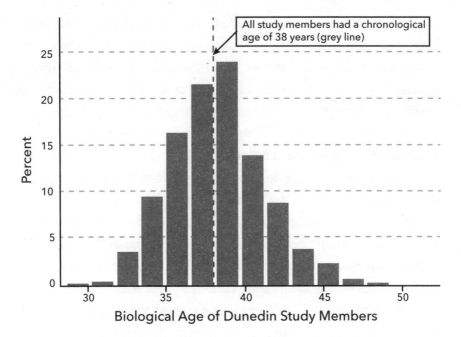

Participants with a chronological age of 38 years in the Dunedin Study showing the spread of biological ages from 28 to almost 50 years.

What was the reason for this almost 22-year variation in biological ageing, even as early as 38? The big driver was low mood and stress, particularly in childhood but also in participants' twenties and thirties.

Furthermore, Belsky and Moffitt tested the hypothesis that those who were biologically older than their chronological age of 38 were continuing to age faster than peers of the same chronological age who retained 'younger' physiologies. They found that a 38-year-old with a biological age of 40 aged 1.2 years faster over the course of 12 years compared with a peer whose chronological age and biological age was 38. In other words, individuals who were biologically older at the first time of data collection continued to age at a faster pace during subsequent years. Furthermore, the pace of physiological deterioration was evident across multiple organ systems: the lungs, mouth, gums and teeth, heart rate and blood pressure, kidneys, liver, eyes, immune function, bone, blood lipids, diabetes markers, body mass index, body fat and the brain. In faster agers, all organs were ageing faster; accelerated ageing was not just confined to one system, it was universal. This suggests that a common mechanism explains biological ageing. If we can pin down this mechanism we could have the key to the elixir of youth.

Already, before midlife, the young adults who were ageing more rapidly were also less physically able. For example, they had poorer balance, unable to stand on one leg for as long as slow agers; had clumsier fine-motor skills, when tested by placing small objects into holes on a pegboard, and had weaker grip strength.

Although young adults were disease free at the time of testing, the test results exposed problems in systems that would ultimately lead to age-related disease – for example, the eyes. The eye is the window to the brain. Small blood vessels in the eye originate from the same source as the small vessels that go to the brain. This shared starting point enables us to draw conclusions about brain vessels from eye vessels in adults. Changes detected in retinal photographs predict future stroke and vascular dementia. Young adults in the Dunedin Study who had older biological age had significantly 'older' eye vessels, thus raising the possibility that they were at a higher risk of stroke and dementia later on in life.

In a parallel experiment, undergraduate students who did not know the study participants or their details were asked to rate facial photographs of them. The students were able to accurately identify differences in the facial ageing of the study members which mirrored biological ageing – they picked out the faster agers as looking 'older'. The faster agers also said they felt older and perceived themselves to be in a poorer state of health.

These findings tell us a few important facts, such as that ageing starts early and that it affects most body systems at once. Why did some 38-year-olds act, look and feel older? The almost 12-year biological age difference was due, in the main, to adverse experiences in youth. However, it is not all doom and gloom. The factors which speed up the epigenetic clock are all modifiable – they are within our control and by doing something about the circumstances that trigger epigenetic ageing, we can influence it at all stages in life. It is never too late to change, though the earlier change occurs, the better. Moreover, not all 38-year-olds who said they were subject to low mood or stress experienced accelerated ageing. Many were resilient to the psychological factors affecting biological change. Noteworthy is that, in the main, these resilient participants had positive perceptions, positive attitudes and optimism despite adverse circumstances.

Perceptions of ageing, feelings of control and emotional responses to getting older are all important factors. This brings us full circle back to my 85-year-old patient and to Eileen Ash, both of whom had positive perceptions, positive attitudes, self-esteem and optimism. It has been argued that perception affects how we age because people who 'feel their age' have diseases or disorders that accelerate ageing and colour their perceptions. But a number of studies from our group and others have now confirmed that we are 'as young as we feel', independent of disease status. In other words, perceptions can override other factors which might otherwise limit physical ageing. Simply feeling younger than one's chronological age slows the pace of ageing, irrespective of diseases or disorders. This is because a positive attitude towards ageing changes cell chemicals in a beneficial manner, possibly by reducing inflammation within the cell and thereby changing the methylation status and epigenetics of the cell. One of our studies showed that people who feel close to or at their chronological age are more likely to develop physical frailty and poor brain health in subsequent years than people who claim to feel younger than their chronological age. This was still the case when we adjusted the analyses to take into consideration any illness or disease at the beginning of the study. When perceptions are negative, they result in declines in self-confidence, self-esteem and satisfaction with life, as well as declines in physical and brain health. Negative perceptions make it more likely that someone will experience diseases such as heart disease, suffer heart attacks in later life and die early.

This brings me to the important influence that language, media and attitudes of friends, family and society have on how we perceive ourselves and how challenging it can be to remain resilient in the face of negative stereotypes. If someone or something continually tells you that you are old, it is hard not to feel old.

Researchers from Yale University showed how quickly ageing perceptions can change an individual's physiology and how the changes become embedded and chronic with exposure to repeated negative stereotypes. In their experiments, adults were exposed to a series of words to describe ageing. For positive stereotypes, these were terms such as 'accomplished', 'advise', 'alert', 'astute', 'creative', 'enlightened', 'guidance', 'improving', 'insightful', 'learned' and 'sage'. The negative stereotypes included 'Alzheimer's', 'confused', 'decline', 'decrepit', 'dementia', 'dependent', 'diseases', 'dying', 'forgets', 'incompetent', 'misplaces' and 'senile'. Participants underwent both mathematical and vocabulary tests that put them under stress after exposure to the stereotypes, while a number of physiological tests were performed to ascertain the biological impact of the mathematical and vocabulary stress tests.

Those exposed to the negative stereotypes demonstrated unwanted excessive physiological responses, with higher blood pressure, higher heart rate and reduced skin blood flow, thereby showing that negative ageing stereotypes rendered the participant less able to mitigate the stress responses. The influence of positive ageing stereotypes, on the other hand, led to more modest physiological responses to stress. In other words, positive stereotypes helped participants deal with stress.

In another research study by our group, adults aged 50 and older were asked to say how much they agreed with 17 statements, such as, 'I have no control over how getting older affects my social life'; 'As I get older, I can take part in fewer activities' or 'As I get older, I get wiser'; 'As I get older, there is much I can do to maintain my independence'. The more these older adults agreed with the negative statements like the first two, and the less they agreed with the positive statements like the last two, the worse their attitudes towards ageing and the more likely they were to experience accelerated physical and cognitive ageing over the following eight years. For example, negative attitudes towards ageing were associated with a decline in walking speed, worse memory and poor performance on a number of other brain tests. This was true even after we accounted for overall health, medications, mood, life circumstances and many other confounders. In other words, perceptions had an independent influence on the physical and mental health pace of ageing.

Furthermore, the science showed that negative attitudes affected how different health conditions interact. Frail participants with negative attitudes had worse mental abilities compared to participants who were not frail. However, frail participants with positive attitudes had the same level of mental ability as their non-frail peers. So, once more, positive attitudes and positive perceptions were protective, emphasising that we really are as young as we feel. Even if we have health problems, attitude still dominates.

When I shared these data with a retired eminent cardiology colleague he responded that he was convinced of the powerful effect of 'mind over heart' and how much stress and perceptions could even influence heart attacks. He shared the following story: 'In 1980, a patient came to see me privately one afternoon and I quickly learned that he had severe angina. I performed an exercise test that confirmed limited blood supply to the heart muscle. I told him that I thought he should undergo coronary angiography to define the anatomy of the blood vessels to the heart. He was very upset as he did not want any invasive procedure and I had to spend some considerable time to convince him that this was necessary. Finally, he agreed and I got him a bed on the private floor of a London teaching hospital where there was no heart monitoring at the time. I was called at 7am the next morning to be told that the nurses had found him dead in bed. Of course, this could have been the natural progression of his disease but his negativity impressed me and made me feel that this contributed to his sudden death.'

On a somewhat lighter note, it is underestimated how much sexual activity is linked to ageing perceptions. Sexual activity is an important part of life for the majority of couples and strongly linked to quality of life – the sexually active have better quality of life, even in old age. In our research, sexually active older adults have more positive perceptions, are less likely to consider themselves old and less likely to believe that ageing has negative consequences. All of these attitudinal factors contribute to superior quality of life and younger biological age for sexually active couples.

As evidenced by my colleague's patient, older adults with negative attitudes about ageing live 7.5 years less than those with positive attitudes, mostly because of higher rates of heart diseases. Our research confirms the link between ageing perception and death. Because we had collected details of many aspects of life and health as part of our study, we were in a position to show that perceptions independently affect early death. So, our own perceptions of ageing and the influence of society on those perceptions are very important to healthy longer lifespan – how we perceive ourselves as ageing is quite literally a matter of life or death.

As Young as You Feel is a 1951 comedy film starring Marilyn Monroe. When printer John R. Hodges (Monty Woolley) is forced to retire at age 65, due to company policy, he decides to do something about it. Dyeing his hair black, he poses as Harold P. Cleveland, the president of his former employer's parent company, and goes on an inspection tour of his old workplace with the firm's nervous, mystified executives in tow. Afterwards, Hodges complains about the lack of experienced, older employees, causing company president Louis McKinley (Albert Dekker) to change the policy. Hodges delivers a rousing speech about the virtues of the older worker. He receives a standing ovation, the newspapers praise him and even the stock market rises on the optimism generated. When the deception is discovered, Hodges has been so successful in turning the company around that Cleveland offers him a job advising him on public relations but Hodges turns him down. In changing the company's ageist attitude and policy he has achieved what he set out to do and is content with the outcome.

Policy initiatives such as mandatory retirement allow employers to force employees to retire at a certain age, usually 65. Mandatory retirement was widespread in the US in the 1960s and 1970s, and still is common in many European countries. Yet, by an extension of the Age Discrimination in Employment Act in 1978, US Congress outlawed mandatory retirement before the age of 70 and in 1986 abolished it altogether. Retirement was redefined. It was no longer an automatic shift in gears from work to non-work at a set age but rather a voluntary withdrawal from the work force at the age that best suits an individual's abilities, interests and career plans. If only this approach was more widespread.

Many European countries still have mandatory retirement

for workers in the public sector despite a large share of workers wanting greater retirement flexibility. In Japan, 43 per cent of workers aspire to continue working past retirement age, whereas in France, only 15 per cent consider this. Two thirds of EU citizens prefer to combine a part-time job and partial pension than to fully retire. In part, disparities in preferences for flexibility across countries are likely driven by various designs of pension systems. The level of pensions available at different ages and the gains from working longer play an important role in shaping workers' attitudes towards flexibility. For example, limits to the amount that can be earned before pension benefits are cut reduce the incentive to work beyond the official retirement age. Yet individuals are not motivated to work longer solely by financial gain – doing so can improve life satisfaction. Surveys from several European countries and the United States have shown that workers over the age of 45 experience less stress and greater life satisfaction, on average, than younger workers. This holds for full-time workers, voluntary part-time workers and the self-employed.

Being able to choose when to stop working is important and affects both life satisfaction and ageing perceptions. I have witnessed the sadness of colleagues, whose enjoyable work was thriving, suddenly forced to retire. Not only was this miserable for the individuals involved but also a great loss to their institutions and to society. In my view, mandatory retirement is ageist and discriminatory and flexible working choice is a more equitable policy.

Unfortunately, mandatory retirement chimes with other negative societal attitudes towards ageing. Reported age-related stereotypes, commonly found in literature and in the media, depict older adults as physically weak, forgetful, stubborn and selfish, with a widespread consensus about these attributes occurring across different cultures and generations. And yet, according to the World Health

Organization, there is little objective medical or psychological evidence to support these commonly accepted 'truths' about ageing. Only a small minority of older adults are physically, cognitively or mentally impaired. The majority are independent and enjoy a good quality of life, which continues to get better after 50. Moreover, negative attitudes about ageing result in social inequalities.

'He told me I was too old for that'; 'She assumed I couldn't understand because of my age' and 'I was denied employment because of my age' are just some of the examples of everyday ageism that 77 per cent of older adults surveyed in the UK reported experiencing. These negative attitudes extend to our social contacts. A 2018 review by the European Social Survey, taking into account the attitudes of 55,000 people across 28 countries, showed that the UK is riven by intergenerational splits, with half of young and middle-aged adults surveyed admitting they do not have a single friend over 70. Only a third of Portuguese, Swiss and Germans said that they have older friends.

In ageist societies, older adults are consequently more likely to be excluded from social situations and have poorer employment prospects than younger counterparts, making it difficult for individuals to perceive themselves as youthful in this cloud of negative attitudes. Worryingly, in some medical situations, older adults are less likely to receive the same treatment simply because of age.

This was graphically demonstrated during the Covid-19 pandemic when, in anticipation of an overwhelming need for intensive care beds and ventilation, some countries introduced a policy whereby patients over a certain age (in most this was 70 years) would not receive intensive care whereas other countries correctly based these decisions on the likelihood of survival and 'biological health', which is a reasonable approach.

In the UK, the approach was ambiguous. Under the Equality

Act, it is illegal to deny an older person access to healthcare on the basis of age. But a 'frailty' screening test was used across the NHS to determine who should be offered more aggressive treatment, with age making up 50 per cent of the frailty score and thereby biasing the assessment against older patients. Dave Archard, emeritus professor at Queen's University, Belfast, argued that an 'overburdened service is no excuse for discrimination that would result in a cull of older people'. He continued: 'To discriminate between patients in the provision of care on the grounds of age is to send a message about the value of people. Such discrimination publicly expresses the view that older people are of lesser worth or importance than young people. It stigmatises them as second-class citizens.' Catherine Foot, director of evidence at the Centre for Ageing Better agreed. 'Chronological age must never be the principal factor that determines a person's right to care. Medically speaking, it is a poor proxy for a person's capacity to respond well to intensive care and to recover.'

The way we think about, talk about and write about ageing has direct effects on health. Ask yourself: are you ageist? Do any of the stereotypes we have discussed resonate with you? We will all grow older and if negative attitudes towards ageing are carried throughout life they will have measurable, detrimental effects on how people age, on how *we* age. If we want a more equitable society now and in the future it is incumbent on all of us to ensure that we do not hold ageist attitudes either towards our own ageing or that of others.

All areas of society need to be aware of the danger of subscribing to negative attitudes. The media makers can take steps to avoid biased language when referring to age. Practitioners should check themselves for bias in treatment strategies. Researchers and policy makers should work together to encourage novel ways to reinforce positive attitudes. The good news is that change is coming because of the sheer number of people who are 'coming of age' and will demand societal equality. Enter the baby boomers, whose attitude towards ageing is different from their predecessors.

Baby boomer is a term used to describe a person born between 1946 and 1964. The baby boomer generation makes up a substantial portion of the world's population, especially in developed countries. After the end of the Second World War, birth rates across the world spiked and the explosion of new infants became known as the baby boom. During the boom, almost 77 million babies were born in the United States alone. Early baby boomers lived to 63 on average whereas later baby boomers can expect to live to 79. Thus, the high numbers coupled with the natural extension in lifespan means that baby boomers represent an influential growing ageing cohort. A large percentage of baby boomers will live up to 25 years longer than their parents did. Those retiring in their sixties can expect to live another 25 years, at least. This is the generation of Woodstock, flower power, hippies, greater access to education, liberal movements and new music genres. Their voices *will be* heard. Baby boomers have high expectations. With more wealth, better health and more energy, and their children now adults, boomers are more likely to be able to afford to spend their retirement fulfilling travel dreams and other bucket-list items. When they reach retirement age, boomers are increasingly likely to be healthy enough to run marathons, build houses and even start new businesses.

◆

When it comes to examining how different countries and cultures address ageing, we need look no further than Denmark. As a society, it behoves us to be aware that ageist attitudes infiltrate our very biology and that childhood circumstances feed into adult health and well-being and thus can have long lasting influences.

Therefore, to achieve a more egalitarian society, childhood and old age are a priority and Denmark exemplifies such a society.

The Social Progress Index evaluates the capacity of a society to meet the basic human needs of its citizens. It is based on an index of social and environmental indicators that determine the quality of life for that country. In short, the index calculates overall human well-being. It includes data from 128 countries on 50 indicators and, extraordinarily, Denmark has most consistently topped Europe's happiness rankings for the past 40 years. Danish society makes it easy to live an interesting, fulfilled life, where age is respected.

Danes spend more money per capita than almost any other nation on children and on older people. Young people get an excellent education and healthcare. Equipped with a strong liberal arts education, Danes make productive employees. Adults spend little time worrying about retirement and focus more on pursuing the jobs they love and can enjoy their final years with the knowledge that the necessities will be covered. It's a virtuous circle.

Danes employ an 'ageing in place' policy. Over 30 years ago, they started to close down nursing homes and to redeploy funds and human resources to enable people to stay in their own homes with support for health needs as necessary. The number of nursing home residents is therefore less than one tenth of that in Ireland, although the population differences are 5.3 versus 4.4 million respectively. For the small number of people who are in nursing homes, the care is delivered in houses with four or five 'apartments' and a central nursing pod. Couples can be together and if one dies, their partner can remain in the nursing home apartment – it is 'home'.

There is a brand of happiness typified in the purpose-driven life of Danes. Like all forms of happiness, it assumes basic needs are covered so that people can pursue their passions at work and leisure irrespective of age. Academics refer to this as eudaimonic happiness, a term that comes from the Ancient Greek word for 'happy'. Worldwide polling company Gallup measures this by asking respondents whether they 'learned or did something interesting yesterday'. The concept was made popular by Aristotle, who believed that true happiness came only from achieving a life of meaning, of doing what was worth doing.

Winters are long and dark in Denmark; night falls by 4:45 on November afternoons. To compensate, Danes create cosy environments with candles, the warmth of the hearth and the gathering of friends of all ages. Ageism is uncommon in Denmark, as is discrimination on any level. The Danes illustrate that it is possible to achieve a more egalitarian society with eudaimonic happiness at each life phase. As a consequence, life expectancy is one of the highest in the world and consistently growing year on year by 0.18 per cent per year – currently 81.11 years.

This egalitarian approach is also apparent in the world's Blue Zones, where generations have respect for each other and friendships transcend age, societal positions and pleasure; friendships and happiness are important for all, independent of age.

———◆———

Our language matters and ageism is exemplified by language and terminology. The terms 'senile', 'demented' and 'aged' are unfortunate terms that are thankfully disappearing. And yet, there is still a term in common use that needs to go: 'elderly'. Certain words can be convenient but promote stereotyping through their generalisation and lack of specificity. Thus, using the term 'elderly' for a person who is robust and independent as well as for a person who is frail and dependent says little about the individual and is an inaccurate and misleading descriptor.

Consider how often you heard references to 'elderly' persons or 'the elderly' during the recent Covid crisis. The terms are ageist. Ageism, like racism and sexism, is a form of prejudice or prejudgement that shapes perceptions. Ageist terms diminish older adults, yet ageism is rampant, even in healthcare, stereotyping older people as sick, frail and physically dependent. Ageism results in less care, less robust care and negatively affects outcomes.

Older adults do not like the term 'elderly' applied to themselves, even if they use the term to describe someone else, as was the case when my 85-year-old patient talked about her 74-year-old 'elderly' neighbour! In a European survey, older individuals displayed a preference for 'older' or 'senior' and strongly rejected the terms 'aged', 'old' and – most strongly – 'elderly'. In 1995, the United Nations Committee on Economic Social and Cultural Rights of Older Persons rejected the term 'elderly' in preference for the term 'older persons'. Additionally, a media guide issued by the International Longevity Centre recommended the term 'older adults' over 'senior' and 'elderly'. The report states, 'After all, we don't refer to people under 50 as "junior citizens".' It is time for our language use to mature: using terms that are precise, accurate, value-free and that older adults prefer makes good sense.

—◆—

I would like to finish where I started, with the science of positive attitudes and successful ageing. The Nun study is a good illustration of how attitudes have far-reaching influence late into the lifespan.

Can you imagine being invited to take part in a study where the researcher asks if you would not only be willing to be examined in detail at regular intervals but also donate your brain to be dissected after you're gone? That is exactly what was asked of the 678 sisters of School Sisters of Notre Dame in the USA, who agreed to take part in David Snowdon's longitudinal study in 1991. The sisters were studied with repeated health tests and psychological tests until their demise. All sisters agreed to have brain pathology studies at post-mortem. In this way, the influences of lifelong health and life experiences on the brain were mapped.

The Nun study was as close as you can get to a long-term human experiment. When conducting experiments, it's important to control as many factors as possible, so as to better investigate the element in which you are interested – in this case brain health and dementia. In this sense, the nuns were perfect: their marital status was the same and none had children, and almost all had been teachers throughout their lives. They had similar income and socio-economic status, ate regularised diets, lived together in

Photographs from the convent archives show reception class from the SSND, in 1927, and the survivors of that class 60 years later (overleaf).

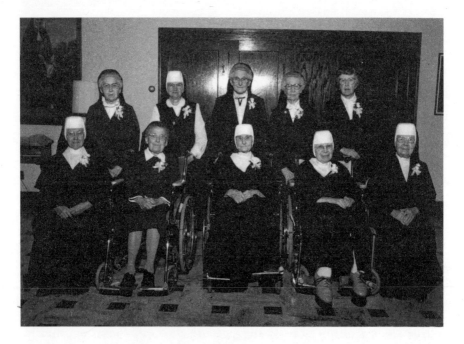

similar surroundings, did not smoke or drink and all had access to similar preventive, nursing, and other medical care services. They rose and went to bed at the same times. In other words, the physical backgrounds and conditions which normally confuse and complicate interpretation of data were about as controlled as they could be.

Some of the unexpected factors that affected whether or not the nuns developed dementia were their attitude in early life and their temperaments. An exciting feature available to researchers was an archived letter that all nuns wrote when they were 20, after their candidate year and before taking final vows. This gave an insight into the attitudes of the nuns and how these attitudes affected the ageing process 60 years later.

Temperament determines people's capacity for coping with stress and life challenges. As illustrated by the Dunedin study, coping mechanisms such as a positive attitude and good temperament help to better manage stress. Positive attitudes provide a type of inoculation to brain pathology. Here are two examples of different attitudes apparent in the candidate year letters:

Sister 1 (low positive emotion): *I was born on September 26, 1909, the eldest of seven children, five girls and two boys. My candidate year was spent in the Motherhouse, teaching Chemistry and Second Year Latin at Notre Dame Institute. With God's grace, I intend to do my best for our Order, for the spread of religion and for my personal sanctification.*

Sister 2 (high positive emotion): *God started my life off well by bestowing upon me a grace of inestimable value. The past year which I have spent as a candidate studying at Notre Dame College has been a very happy one. Now I look forward with eager joy to receiving the Holy Habit of Our Lady and to a life of union with Love Divine.*

Simply put, the nuns who expressed more positive emotions lived, on average, a decade longer than their less optimistic peers and were less likely to get dementia. By the age of 80, 60 per cent of the least happy nuns had died. The probability of survival was consistently in favour of the more positive nuns.

How we perceive ourselves influences the biological pace of ageing. Perceptions can be influenced by societal attitudes, ageism and life course experiences. The more optimistic and positive our perceptions, the more likely that we will live longer, healthier and happier lives. This is explained by changes in biological ageing, as evidenced by the DNA methylation in cells throughout the body. I hope that this awareness will empower individuals to age in the most successful way possible and even enable our last decades to be our best quality years.

Chapter 2

Why Do We Age?

IN 25 YEARS, EVERY FOURTH PERSON LIVING IN EUROPE AND North America will be aged 65 or over. The biggest increase will be in the over eighties. The number of persons 80 years or over is projected to triple, from 143 million in 2019 to 426 million in 2050. In 2018, for the first time in history, persons aged over 65 outnumbered children under 5 worldwide.

There are areas of the world where both men and women lead extraordinarily long lives and where the proportion of people living to 100 years and beyond is larger than elsewhere. These areas are known as the Blue Zones.

The concept of the Blue Zones grew out of studies published in 2004, when social biologists Gianni Pes and Michel Poulain identified a province in Sardinia as the region with the highest concentration of centenarians. As the two men zeroed in on the cluster of villages with the highest longevity, they drew thick concentric blue circles on the map and began referring to the area inside the circle as the 'Blue Zone', thus cementing the title in scientific and public parlance. Dan Buettner is a journalist who, although he had no specific expertise in science or gerontology, became interested in Poulain's Blue Zone studies. Together with Pes and Poulain, the term 'Blue Zone' was broadened by Buettner to apply to other validated longevity areas: Okinawa, a Japanese island in the Pacific Ocean; the Seventh-day Adventists community in Loma Linda (Spanish for 'beautiful hill') in San Bernardino, California; Nicoya, a peninsula on the Pacific coast of Costa Rica and Ikaria, a Greek island and small archipelago in the Aegean Sea. This broader concept was published in *National Geographic* in 2005 and became one of the most highly cited articles ever published by the magazine. Based on data and first-hand observations of life in the zones, scientists began to offer explanations for why these populations live healthier and longer lives. Their studies underpin today's understanding of longevity.

Remarkably, all of the peoples in the Blue Zones share similar lifestyle characteristics, despite the fact that the zones are many miles and even continents apart. The factor of paramount importance is that ample physical activity, such as walking, gardening and housework, is embedded in daily routines. For Blue Zone centenarians, exercise is not a fixed purposeful action like a gym session or class but rather built into every opportunity. At a recent lecture I attended, Poulain showed a striking video of a woman in her late nineties chopping firewood, something she did every morning of her adult life.

Another feature of Blue Zone centenarians' lives is having a purpose. The Okinawans have a special name for 'purpose', they call it *ikigai*, while the Nicoyans call it *plan de vida* – knowing when you wake up in the morning what your plans and achievements of that day will be. Subsequent research has established that having a purpose in life makes us healthier, happier and, amazingly, can add up to seven years of extra life. Having a sense of belonging and strong, close family connections with spouses, parents, grandparents and grandchildren contributes to the sense of purpose and is commonplace for Blue Zone centenarians' life histories. In the case of the Adventists, their 'purpose' is being part of a faith-based community, which adds 4 to 14 years to life expectancy.

Whereas stress is part of everyday life for all of us, some of us struggle more than others to cope with it. Blue Zone centenarians have longstanding stress-relieving rituals built into daily routines. Ikarians take an afternoon nap; Sardinians do 'happy hour', sharing wine and conversation with friends and family; Adventists have communal prayers – all of which are 'downtime' destressing activities. Relaxation, social engagement, laughter, friendship and meditation are part of many of their lifelong rituals. Later on, I will explain how downtime benefits the nervous and cardiovascular systems and slows the biology of ageing.

Diet is an important contributory factor to longevity. I know from experience that people glaze over when a professional speaks about food and diet. I recall doing a radio interview and, as I turned to the subject of diet, the presenter retorted, 'Oh, not that boring old chestnut again!' Many scientists maintain that food is the key to ageing well. What is special about the Blue Zone diets is how similar they are, despite the physical and cultural distances between the communities. The diet is predominantly plant-based: beans are its cornerstone and vegetables, fruit and whole grains round it out, with meat eaten only in small amounts. When eating, Blue Zones centenarians apply an 80 per cent rule – that is, they stop when the stomach is 80 per cent full and eat the smallest meal in the early evening. How far removed this is from the chant that certainly I and I expect some of you were familiar with when growing up: 'Finish up everything on your plate; there are too many people starving in the world.' Even as a child, that made no sense to me.

Below is a list of lifestyle behaviours shared by Blue Zone centenarians that determine longer, healthy lives:

1. Life purpose
2. Stress reduction
3. Moderate caloric intake
4. Plant-based diet – semi-vegetarian
5. Moderate alcohol intake, especially wine
6. Engagement in spirituality or religion
7. Engagement in family life
8. Engagement in social life
9. Regular physical activity

It is worth emphasising that older people in the Blue Zones not only have extended longevity but also remarkably good health, with fewer illnesses in old age than we encounter elsewhere. The Blue Zone people therefore have the ideal scenario – better health

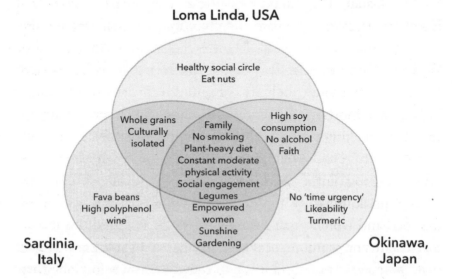

Overlap in healthy behaviours in three of the Blue Zones

coupled with longer lives. This is attributable to a delicate balance between the traditional lifestyle still practised in the areas and some modernity, with increased wealth and better medical care. But another key factor is happiness. By and large, Blue Zone centenarians are a happy lot with positive dispositions.

—◆—

You can well imagine that one of the challenges for research into regions and areas of proposed longevity is verification of the age of so called 'long-lived' persons. After all, we have all probably lied about our age at one time or another. How can we be sure that a claimed age is accurate, even when 'records' of birth are available? The fact is that falsifications of records do happen. The following is an illustration of this challenge.

In the January 1973 issue of *National Geographic*, the physician Alexander Leaf gave a detailed account of his journeys to communities of purported long-living people: the Hunza from Pakistan, the Abkhazians from the Soviet Union and Ecuadorians from Vilcabamba. According to Leaf, there were ten times more centenarians in these communities than in most Western countries. This was despite the fact that, as he pointed out, each of the countries had poor sanitation, infectious diseases, high infant mortality, illiteracy and a lack of modern medical care, thus making the inhabitants' extreme longevity even more extraordinary. However, unfortunately for Leaf, who I am sure acted in good faith, some years later, 'age exaggeration' was proven, predominantly in Vilcabamba, where a large number of the men and women had inflated their age in order to improve social status or promote local tourism. Later, Leaf acknowledged that there was no substantive objective evidence for longevity in the village of Vilcabamba. Further studies confirmed that none of the aforementioned areas survived scrutiny. In light of

Leaf's experience, Poulain and colleagues conducted rigorous surveillance and validation studies for the Blue Zones, eventually confirming that a much higher proportion of persons lived into very old age with good health and lower levels of arthritis, heart disease, dementia and depression in these regions than elsewhere in the world. The observations and details of the Blue Zones withstood scrutiny.

At the time of writing, the record-holder for longevity for the human species, at 122 years and 164 days, is a French woman, Jeanne Louise Calment. I really like her life story because, for me, it epitomises all of the elements which make for successful ageing. Calment was born in Arles, Bouches-du-Rhône, Provence, on 21 February 1875. Her father was a shipbuilder who lived to 93; her mother died aged 86. She had an older brother, François, who lived to the age of 97 – a strong family history of longevity. At 21, she married an heir to a drapery business and the couple moved into a spacious apartment above the family store in Arles. Jeanne never had to work, employed servants and led a leisurely lifestyle within upper society, pursuing hobbies such as fencing, cycling, tennis, swimming, roller skating, playing the piano and making music with friends. In the summer, the couple would mountaineer. She enjoyed an idyllic, relatively stress-free and fun-filled life, devoid of monetary worries with much variety, pleasurable activities and exercise. She had one daughter who died of pleurisy aged 36. Her husband died aged 73, reportedly from cherry poisoning.

In 1965, aged 90 and with no heirs left, Calment signed a life estate contract on her apartment to André-François Raffray, selling the property in exchange for a right of occupancy and a monthly revenue of 2,500 francs (€380) until her death.

Raffray died 30 years later, by which time Calment had received more than double the apartment's value, and his family had to continue making payments to her. Calment commented on the situation by saying, 'In life, one sometimes makes bad deals.' In 1985, she moved into a nursing home, having lived on her own until age 110.

In the nursing home, Calment initially followed a rigorous daily routine. She was woken at 6:45am and started the day with a long prayer at her window, thanking God for being alive and for the beautiful day that was starting, underscoring her positive attitude and outlook. Seated on her armchair, she did gymnastics wearing her stereo headset. Her exercises included flexing and extending the arms, hands and then the legs. Nurses noted that she moved faster than other residents who were 30 years younger. It has since been well established that an individual's walking speed is a strong predictor of healthy longevity. Her breakfast consisted of coffee with milk and rusks.

She washed herself unassisted with a flannel cloth rather than taking a shower, applying first soap, then olive oil and powder to her face. She washed her own glass and cutlery before proceeding to lunch. She made herself daily fruit salads with bananas and oranges. She enjoyed chocolate and, after the meal, smoked a Dunhill cigarette and drank a small glass of port. At this juncture, I'll share that my husband, who is partial to both port and a cigar, frequently cites Calment as an example for why his habit may be beneficial rather than harmful, usually when I'm mid-sentence admonishing him for the cigar! In the afternoon, she would take a nap for two hours and then visit her neighbours in the nursing home, telling them about the latest news she had heard on the radio. At nightfall, she would dine quickly, return to her room, listen to music (her poor eyesight due to cataracts, which she refused to have surgery for, prevented her from enjoying her crosswords pastime), smoke a last cigarette and go to bed at 10:00pm. On Sundays, she went to Mass and on Fridays, to Vespers.

Apart from aspirin for migraines she had never taken medicine, not even herbal teas. She did not have hypertension or diabetes and her blood sample analytics in her last year were in normal ranges. Unfortunately, aged 114 she had a fall, sustained a hip fracture and was in a wheelchair thereafter, but lived for almost another nine years. Calment remained 'mentally sharp' until the end of her life. A documentary film about her life, entitled *Beyond 120 Years with Jeanne Calment*, was released in 1995.

But even Calment's story did not escape the scrutiny and questioning of the scientific community. In December 2018, Jeanne Calment's record of longevity was contested by a Russian gerontologist, Valery Novoselov, who was an assistant professor – the first step of the academic ladder – at the University of Moscow, and by Nikolay Zak, a laboratory technician. Their sceptical paper was published on a website, not a peer-reviewed journal, and Calment's story was also challenged by them in a manuscript posted on ResearchGate.net. The sceptics proposed a fraudulent conspiracy by her family to identity switch between mother and daughter. Furthermore, they contested that it was mathematically impossible to live to Jeanne Calment's advanced age. Despite the total absence of validation for their claim, and the absence of peer review of their assertions, it created quite a storm in the media and gerontology community. I recall having dinner with a renowned UK gerontologist the night before the claim was made public. He told me with some glee of this amazing story which would 'break tomorrow and discredit and defame Jeanne Louise Calment and her family'. Even he never thought to question its validity!

But the claim was incorrect and was followed up a year later with a verified rebuttal and presentation of full details confirming Jeanne Calment's age in a proper peer-reviewed paper, thus discrediting Zak and Novoselov.

Calment's life exemplifies many of the features of lifestyle in the Blue Zones and covers all of the secrets for successful ageing. She had financial security, was relatively free from stress, experienced a varied and full life, had plenty of outdoor activities, was consistently curious throughout her life, had lots of friends and social engagement, ate good food and practised beneficial routines and rituals until her death. Had André-François Raffray known how her family history and lifestyle would contribute to healthy, longer living, he may well have declined the property deal in 1965, which cost him dearly! Most of us who read Calment's story will surmise that her longevity was thanks to 'great genes' but there are a number of theories worth discussion.

Early researchers concluded that the process of ageing was linked to fertility. In other words, they thought that mortality increases as fertility declines and all biology is dependent on this relationship. Although many species, like humans, do show mortality trajectories consistent with this theory, there are plenty of violations. In some species, such as the desert tortoise, mortality declines with age, while in others, such as the tiny freshwater organism known as the hydra, it remains constant. So fertility does not explain why all animals age. Moreover, and amazingly, the shapes of mortality trajectories are not strongly associated with the lifespan of species. In other words, both short- and long-lived species have increasing, decreasing or constant mortality rates. For example, humans and other mammals are more likely to die with increasing age, while plants are highly variable.

Longevity can be manipulated. Probably one of the greatest breakthroughs in the field so far is manipulation of genes and thus longevity and its plasticity, i.e., slowing down or speeding up ageing. We know, for instance, that manipulation of DNA repair systems in mice sometimes accelerates ageing. On the other hand, we can turn off a single gene, like growth hormone receptor gene, and significantly extend the life of a mouse. Much research has been invested in this very approach – switching genes 'off and on' – in order to reduce diseases and slow down the ageing process. To date, these studies have only been conducted in animals and are not as yet safe to conduct in humans. But there are other theories for why we age apart from genes. Knowing the various theories will help understanding of what we as individuals can do to slow down ageing and achieve a healthy lifespan close to that of the Blue Zone people.

———◆———

There are a number of explanations as to why, apart from genetics, cells age – and no one explanation clearly dominates. One theory is that an accumulation of toxins, free radicals and bad proteins builds up in cells, thus causing damage which eventually kills the cell. Another is that ageing is programmed, i.e., through an internal clock to live to a certain age. A more recent popular theory is that the immune system alters as we grow older, 'attacking' and eventually killing us.

We will briefly explore each possibility because they inform the recommendations for successful ageing in subsequent chapters. I have tried to simplify the explanation of the science and have kept it as brief and as straightforward as possible.

Let's first start with genes because, in my experience, this is the theory that is most strongly embedded in popular belief. Genes are responsible for up to 30 per cent of how long we live for up to age 80 and play a much larger role in likelihood of longevity thereafter. Even very recently, I gently disabused a patient of his firm conviction that 'genes are everything' and that he had nothing to worry about because his father had died aged 94 and his mother, aged 87. Therefore, he, aged 68, assured me that although he smoked 20 cigarettes per day, was overweight and consistently enjoyed at least a half a bottle of wine daily, it would not make any difference to his health. After all, he said smiling, 'I have great genes.' His pronouncements are not entirely correct. Ageing is, only in part, determined by the genes we inherit.

We have two copies of each gene – one inherited from each parent. Most genes are the same in all of us but a small number (less than 1 per cent) are slightly different between people. We have between 20,000 and 25,000 genes. Allelles are forms of the same gene with small differences in DNA. These small differences make up our unique physical features.

'Twin research' has taught us a lot about genes and ageing. Identical twins provide a 'natural experiment' because they have the same genes at birth and so are 'genetically programmed' to age in the same way – however, they do not do so! This is because life experiences and environmental factors including lifestyle behaviours (such as my patient's smoking, drinking and diet) have a big effect on how fast or slowly we age and predominantly determine how long we live for.

An early study of 2,872 Danish identical twin pairs compared the relative contribution of genetics and other 'environmental' factors. Born between 1870 and 1900, the genetic influences in the twins in this study were minimal up to late adult life but did get stronger after this. In other words, childhood experiences, social and economic circumstances, marital status, diets, sleep, smoking habits, alcohol intake, depression, stress and physical activity predominantly influenced ageing for the first numbers of decades, with genes only becoming more dominant in latter years. Other twin studies followed, confirming that genes contribute to only 20–30 per cent of the variation in survival up to 80 years and genetic factors play a much stronger role in long living beyond 80. The remaining 70–80 per cent of variation in survival up to age 80 is due to external or environmental factors. So, if my patient makes it to 80, then his hypothesis about his genes protecting him may prove correct. He is much more likely to succumb to problems from his lifestyle behaviours well before that.

So genes play a dominant role in exceptional longevity – that is, reaching 100 or more. But exceptional longevity is a rare trait – in a sample of 5,000 in the US, only 1 person will be a centenarian; only 1 in seven million is a super-centenarian (age 110+ years). Impressively, siblings of centenarians are more likely to live to 100 than others born in the same year. Therefore, genes play a big role in super-ageing and we have already identified some of these genes, such as DAF2. Many of the genes associated with exceptional longevity are involved in regulation of blood sugar and food metabolism, and in the cell's energy production and metabolic rate. You will appreciate the enthusiasm to understand whether we can manipulate these genes to try to reduce the frequency of health problems in latter years for all of us.

But let's return to 'environmental' factors and ageing. You will be aware of the expression 'you wear your heart on your sleeve'. Well, when it comes to ageing, you wear your age on your face! The ageing face is a good illustration of our cells ageing. Facial skin cells and tissues exhibit all of the hallmarks of ageing and are there for all to see. My mother believed that you could tell a smoker by their skin because smoking speeds up the pace of ageing. Another recent twin experiment has proven her to be correct. Researchers in Ohio recruited nearly 200 sets of identical twins who were attending an annual twin festival. Using photographs of each twin set, the researchers asked an independent panel to rate differences in the twins' appearances, to decide whether they thought one twin looked older than the other and to guess their age. They found several factors influenced appearance and facial ageing, including smoking and excessive sun exposure – 10 years of smoking added 2.5 additional years of ageing to a twin's face, compared with a twin who didn't smoke.

Stress also influenced the panel's interpretation of the photos: divorced twins looked two years older on average than a twin who was married or widowed. Twins taking antidepressants also looked older – possibly because depression itself increased facial ageing or because use of antidepressant drugs relaxed facial muscles in a way that increased the appearance of ageing. Facial ageing and body weight were also linked. A heavier body weight before the age of 40 was associated with an older appearance. However, in women over 40, a heavier body weight was associated with a more youthful look, compared to a thinner twin. I recall listening to an interview with the actress Kathleen Turner over a decade ago in which she claimed that 'after a certain age' we should sacrifice a bit on the hips for the sake of the face – a view that seems to be backed up by this research. So lots of external factors other than genes contributed to an older appearance in identical twins.

———◆———

Every cell has one nucleus. The nucleus is the cell's 'library' and gives instructions for all of the cell's activities, including everything that regulates ageing. The nucleus houses our chromosomes, which hold our genes and therefore our DNA, dictating everything that we are. DNA is responsible for the division of our cells throughout life. Each cell has 46 chromosomes, made up of protein and a single molecule of DNA. Our liver cells only use the 'liver DNA', the rest is switched off. Our eyes only use the 'eye DNA', etc.

At each end of a chromosome is a telomere, often compared to the plastic tip at the ends of a shoelace. Telomeres are a hot topic in gerontological science because they protect the chromosomes, preventing them from unravelling, sticking to each other or changing shape. Damaged chromosomes cannot efficiently send messages from the nucleus to other cell structures. Each time a cell divides (replication), the DNA separates in order for the genetic information to be copied. When this happens, the DNA coding is duplicated – with the exception of the telomere, which is not. When the copy is complete, the copy separates from the original at the telomere. As such, with each cell division, the telomere gets shorter and shorter until it can no longer completely protect the chromosome. It is then that the cell dies. We use the length of a telomere to determine the age of a cell and how many more divisions remain for that cell. Hence the gerontological interest in telomeres.

Ageing is characterised by break-up of sections of the chromosomes in the nucleus, which disrupts transfer of vital information from the nucleus or 'library' to the rest of the cell. Therefore, the instructions from the nucleus become flawed. These instructions include information on replication of the cell, production of energy and removal of waste materials. Flawed information results in slow functioning, inefficient operation and, eventually, the death of a cell.

In the end, all of our cells are 'mortal' with the exception of one cell type – cancer cells. Unlike normal cells, cancer cells do not undergo programmed cell death but continue to multiply without end. Therefore, they eventually take over all other cells and body organs, this is what we know as metastases. Cancer cells show no telomere shortening which may be the very reason for their survival. A better understanding of telomere survival in cancer cells may help us to manipulate telomere shortening in normal cells and thereby delay ageing. At present, we cannot manipulate human genes or human telomere length. But it's a different story for mice genes.

It is possible for scientists to manipulate the breakdown of chromosomes in mice to make cells younger. This discovery won the Nobel Prize in Physiology or Medicine for Shinya Yamanaka in 2012. He was able to turn mature cells into young cells that

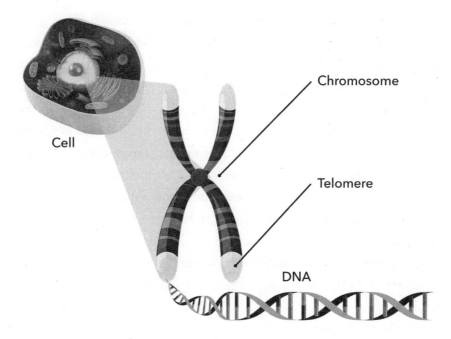

had the capacity to change into a number of different cell types – these cells are known as pluripotent cells. The early human embryo consists mainly of these pluripotent cells, which may become a nerve cell, skin cell, heart cell or liver cell and start the growth thereafter of that organ-system in the embryo. Shinya Yamanaka succeeded in identifying a small number of genes in mice that regulate transition from mature to pluripotent cells. When these genes were 'switched on', skin cells could be reprogrammed as immature pluripotent cells, meaning they could grow into the cell type of the scientists' choosing. This major discovery holds great potential for future manipulation of ageing as well as for developing new approaches to organ transplants.

Proteins that function as recycling trucks to take the waste and toxins from within cells to recycling centres in the cell and beyond are also switched on and off by instructions from the nucleus. In animals genetically engineered to produce higher levels of these proteins, lifespan is extended by 30 per cent. This is remarkable.

The average male in the Western world lives to 80. Should we be able to manipulate these 'recycling' proteins, lifespan would increase such that the average would be 105. The oldest man in the UK is 111 – he could be 141 if we were able to manipulate these proteins. Many age-related diseases occur because of an inability to clear the waste fast enough through the cells by the recycling trucks – diseases such as arthritis, heart disease, cancer and dementia. The clearance process that cells use to destroy and recycle cellular waste is known as autophagy. Research in this area won Yoshinori Ohsumi, a Japanese cell biologist, the 2016 Nobel Prize in Physiology or Medicine. Ohsumi discovered how autophagy works and its relevance to ageing. Work is now focusing on ways to manipulate autophagy in order to extend healthy lifespan.

———◆———

Another theory is that we are programmed to age, that each of us is programmed at birth to die at a certain age and this is down to the genes we inherit. In support of this theory is the lack of variation in lifespan within species. Elephants die at around 70, spider monkeys die at around 25 and humans die at around the age of 80.

Technically, there is really no reason why the human body should age as long as it can repair and renew itself. Were this the case, individuals in the species would just keep on living until an accident or other external event killed them. Yet, as humans age, we experience changes to almost all physiological functions – hormones, the immune system, muscle function, heart function, lung function, blood systems and brain function. Therefore, something other than time must cause ageing. The programmed theory of ageing asserts that ageing is an intentional process, that is we are pre-programmed to age and die.

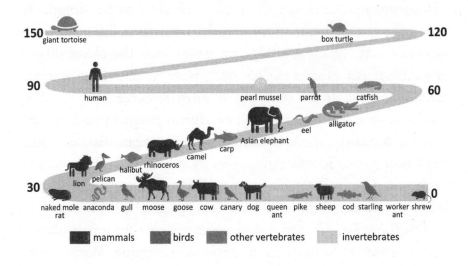

150 giant tortoise   box turtle   120

90   human   pearl mussel   parrot   catfish   60

                              Asian elephant   eel   alligator
                        carp
                 camel
          rhinoceros
     halibut
30   pelican   0
   lion
naked mole anaconda gull moose goose cow canary dog queen pike sheep cod starling worker shrew
   rat   ant   ant

   ■ mammals   ■ birds   ■ other vertebrates   ■ invertebrates

The above diagram illustrates the differences in the average lifespan of different animals. (By permission Silvin Knight 2020. Data from S.S. Flower, 'The Duration of Life in Animals' in Proceedings of the London Zoological Society.)

The rate-of-living theory states that humans and other living organisms have a finite number of breaths, heart beats or other measures, and that they will die once they've used these up. This is an attractive theory for which there is some evidence. There is a clear link between heart rate and lifespan in most animals. Small animals have higher heart rates and shorter lifespans; larger animals have slower heart rates and longer lifespans. There is as yet no definite evidence that humans have a finite number of heart beats, although people with faster resting heart rates die earlier.

The free-radical theory is a well-known and much promoted ageing theory, particularly by corporates with a vested interest in supplements. When cells create energy, they produce unstable oxygen molecules called free radicals. Free radicals are some of the 'waste' products we referred to earlier. The theory posits that excess of free radicals causes accelerated ageing. Antioxidants are substances found in plants that soak up free radicals like sponges.

In laboratory experiments, higher numbers of antioxidants minimise the damage caused by free radicals. However, most human studies of antioxidant supplements have not as yet shown the same dramatic effects. It is not entirely clear why this is the case but I will examine supplements in more detail later.

The protein cross-linking theory attributes ageing to excessive binding between proteins in cells that forms rigid structures like ladders within the cell and causes the structural changes and stiffening characteristic of age-related disorders, such as hardened arteries, cataracts, skin wrinkles and lung fibrosis.

Finally, a popular theory about ageing is that it is mostly due to inflammation occurring because of malfunctioning in our immune systems as we get older. Our immune system fights infection and anything else that is 'foreign' to the body. The effectiveness of the immune system peaks at puberty and gradually declines thereafter. Impaired immune responses lead to cell inflammation and eventually cell death.

Thanks to the global roller coaster that is Covid-19, we have all become very aware of the critical role the immune system plays and how fragile it is, particularly in older ages. Older people are twice as likely to have a serious response to Covid-19 as ageing immune systems make it harder to fight off the infection. Less than 1 per cent of people in their twenties died from the Covid-19 virus, compared with 20 per cent of over 80-year-olds. Overall, 80 per cent of deaths were in people over 65. In Italy, a country with one of the world's oldest populations, the average age of death from Covid-19 was 81. This put a 20-year gap between the average age of people who tested positive for the virus and the deceased. Therefore, the potential to delay or reverse the effects of ageing on the immune system would have significant benefits for current and future infections. Understanding the cell changes that underlie the decline in immune function has advanced in recent years and clinical trials to evaluate various methods by which to enhance immunity are now underway, leaping to prominence because of Covid-19.

◆

In summary, it is most likely that all of the theories detailed above contribute, in one way or another, to cell ageing and cell death. Ageing is unlikely to happen because of a single reason – it is multifactorial. However, the good news is that the various interventions that we will discuss, such as food, hormesis, exercise, sex, laughter, friendship and sleep, all have their effects at the cell level through one or more of the pathways involved in the theories of ageing.

A significant chunk of health budgets in Western countries is drawn down by age-related issues. Some argue that delaying ageing merely yields a one-time benefit after which the same healthcare expenses would ensue. But the evidence is to the contrary. We know from animal experiments that delayed ageing produces a genuine compression of diseases and death rates. In other words, if cell ageing is slowed down, it reduces the period of time at the end of life during which the animal experiences age-related diseases. For example, calorie-restricted animals not only experience a reduction in their risk of death but also declines in the risk of a wide variety of age-related conditions, such as cataracts, kidney diseases, arthritis, dementia, and many others. If this could be achieved in people, the benefits to health and vitality would begin immediately and the costly period of frailty and disability at the end of life would occur over a shorter interval before death. Such compression of diseases and disability would create financial gains, not only because ageing populations have more years to contribute but also because there would be fewer years during which age-entitlement and healthcare programmes were used.

The pace of population ageing around the world occurs at different rates. France had almost 150 years to adapt to a change from 10 per cent to 20 per cent in the proportion of people older than 60. However, Brazil, China and India have had only 20 years to make the same adaptation. This will put huge pressure on health and social care systems in these countries. Thus the world faces major challenges to ensure that health and social systems are ready to address ageing demographics.

The countries in Europe where citizens enjoy the highest rates of good health after age 65 are Sweden and Switzerland. Why? What is different about Sweden and Switzerland? Explanations include better diets, good healthcare, high rates of physical activity and egalitarian societies. In other words, factors that are well within our control as individuals and as a society. Understanding the ageing process will not only inform what we can do as individuals to ensure longer, healthier lives but also how society can better assist its citizens as they age, ensuring equity.

Slowing down the process of ageing by just seven years could cut in half diseases at every single age. This would have a massive impact on human lifespan and healthcare costs. The Wright brothers, who built and flew the world's first successful aircraft, looked at birds and thought, 'These birds are heavier than air and they can fly. If birds can fly, we can make aeroplanes.' And they did. There is no law of nature that says ageing is immutable. So, let's be optimistic about what we can do for ourselves, both now and in light of new discoveries that are close to hand.

Chapter 3

Friendship

OUR FRIENDS AND RELATIONSHIPS LITERALLY KEEP US ALIVE. Even Albert Einstein recognised the influence of friendship: 'However rare true love may be, it is less so than true friendship.' When I started to study the association between family ties, friendship and health in the TILDA study, I was staggered by how powerful the physical effects of friendship are and how much difference good friendships make – and not only to pleasure and quality of life, but to hard outcomes like heart disease and even determining when we die. Good friendships add years to our lives.

I came across a lovely story recently of an unlikely 'pairing' that emphasised why friendship is important not just to us humans but also to other mammals. The story is about a rescue long-tailed macaque and a stray black-and-white kitten. The kitten wandered onto a Wildlife Friends Foundation park in Phetchaburi, Thailand, and was quickly adopted by a resident, Jojo. The macaque monkey had been rescued by the Wildlife Friends some years before when they came upon him cruelly caged in a restaurant. He was kept in the cage alone and was being used as a photo prop to amuse customers. Needless to say, the isolation that Jojo would have experienced is particularly painful for social animals like the macaque that, like ourselves, are a gregarious species, enjoying life in communities and organised groups. However, six years later, Jojo is a whole new monkey; he's become the leader of a group of other rescued macaques at the wildlife rescue centre and has adopted a new best friend, the stray kitten. The furry duo has put species differences aside and share food, pose for pictures and even check each other for lice. The story exemplifies how friendship can unlock unexpected bonds and pleasures across species.

In 2020, governments around the world responded to the Covid-19 crisis by encouraging or mandating isolation for long periods of time. The long-term consequences of this blunt approach to our gregarious populations are yet to be realised. In this chapter, we will explore how isolation is counterintuitive for humans and, just as it was for Jojo, bad for our physical and psychological health.

———◆———

One of the most compelling accounts of the nature and power of friendship is that expressed by the Roman orator Cicero (106–43 BC), who in his classic essay 'On Friendship' ('Dé Amicitiá') wrote:

> 'If the natural love of friends were to be removed from the world, there is no single house, no single state that would go on existing; even agriculture would cease to be. If this seems a bit difficult to understand, we can readily see how great is the power of friendship and love by observing their opponents, enmity and ill will. For what house is so firmly established, what constitution so unshakable that it could not be utterly destroyed by

hatred and internal division. From this may judge how much good there is in friendship.'

Not far from Cicero's birthplace, Arpinum, is the traditional hill town of Roseto Valfortore, which was the scene of one of the most groundbreaking studies on friendship and health as it firmly established the contribution of friendship to biological ageing. The *comune* lies in the Apennine foothills of the Italian province of Foggia. The town is organised around a large central square and the church. Narrow storey steps run up the hillside, flanked by closely clustered two-storey stone houses with red tile roofs. For centuries, the Rosetans worked in the marble quarries in the surrounding hills or cultivated the fields in the terraced valley below, walking four or five miles down the mountain in the morning and then making the long journey back up the hill at night. It was a hard life. The townsfolk were barely literate and desperately poor, and had little hope for economic betterment – until, at the end of the 19th century, word came of a land of opportunity across the ocean.

In January 1882, a group of Rosetans set sail for New York. They eventually secured employment in a slate quarry near Bangor, Pennsylvania. Thereafter, other families followed, joining their compatriots in the slate quarry. Those immigrants, in turn, sent word back to Roseto and by 1894, some 1,200 Rosetans had applied for passports to America, leaving entire streets of their village abandoned. However, when they reached the new land, they re-established their old familiar village, building two- and three-storey houses on the hillside and clearing and cultivating the land, planting onions, beans, potatoes, melons and fruit trees in the long backyards behind their houses. The town came to life. The Rosetans began raising pigs and growing grapes for homemade wine. Schools, a park, a convent and a cemetery were built. Small shops and bakeries and restaurants and bars opened.

Neighbouring Bangor was largely Welsh and English, and the next town over was overwhelmingly German, which meant that – given the fractious relationships between the English and Germans and Italians in those years – new Roseto stayed strictly for Rosetans. The precise southern Foggian dialect was the language spoken. Roseto, Pennsylvania, was its own tiny self-sufficient world, all but unknown by the society around it, and may well have remained so but for Stewart Wolf.

Stewart Wolf was a pioneer of psychosomatic medicine. He was born in 1914 in Baltimore and died of Alzheimer's disease in Oklahoma City in 2005, aged 91. He began studying the people of Roseto, Pennsylvania, in the early 1960s after a doctor from Roseto told him that he hardly ever saw patients with heart attacks who were younger than 50. That was in stark contrast to neighbouring towns and elsewhere in the US, where death from heart attack in men over 40 was of epidemic proportions. The heart attack death rate in Roseto was half the rate of elsewhere in the USA. The statistics confirmed that Roseto was a starkly healthier place to live, and no one could guess why. Wolf thought that there might be something specific about the lifestyle of the town's residents, most of whom were Italian immigrants, that had a beneficial effect on health.

In his publications, Wolf described Roseto, Pennsylvania, as a pretty but modest village with almost 2,000 inhabitants. In 1962, Wolf and his research team descended on Roseto with the full equipment of scientific investigators to establish why there was this remarkable difference in heart attack rates. After years of work, including extensive history taking, detailed physical examinations and blood analyses, the reasons were still not evident. It could not be explained by genetics, because Rosetans from nearby towns were not protected from early cardiac deaths, nor by differences in diet, smoking, exercise or body weight. It was a mystery.

Then, one Sunday, it dawned on Wolf, as he sat in the square watching the Rosetans pour out of the church, loitering, chatting and laughing before returning home for a long lingering lunch with extended families and friends: the secret of Roseto was Rosetans themselves. What was different was the attitudes, friendships, respect for family, constant social contact and sense of fun that pervaded Roseto. In 1964, Wolf and colleagues published a paper in the *Journal of the American Medical Association* on the subject, concluding that it was indeed social interactions with friends and family that accounted for the low rate of heart attacks. With his long-time research associate John Bruhn, a social scientist, he coined the phrase 'the Roseto effect'.

Wolf noted how, often, three generations lived in one household and that there was constant interaction with family, neighbours and the community. There were 22 civic societies for a population of 2,000. Rosetans had firm family ties and friends and retained traditional and cohesive family and community relationships. There was no crime rate and few applications for social assistance. The society was egalitarian. Rosetans, regardless of income and education, expressed themselves in a family-centred social life. There was a total absence of ostentation among the wealthy, meaning that those who had more money didn't flaunt it. There was nearly exclusive patronage of local businesses, despite nearby bigger shops and stores in other towns. The Italians in Roseto intermarried from regional cities in Italy. Families were close knit, self-supportive and independent, but also relied in bad times on the greater community for assistance and friendly help.

No one was alone in Roseto. No one seemed unhappy or stressed out. Nearby wealthier towns suffered almost double the rate of heart disease, though their medical facilities, diet and occupations were either better or at least equal to Roseto's.

In their book, which I highly recommend, *The Power of Clan*, authors Wolf and Bruhn covered the story of the town 1935–1984. They emphasised how residents had avoided the internalisation of stress by sharing resources, worries and emotions. But as intermarriages (Italian to Italian) became less frequent, social ties between family and community were dismantled, wealthy Rosetans took up conspicuous consumption and adopted other modern behaviour, a clear correlation could be shown with an escalation in heart disease. Eventually, the heart disease rates in Roseto were the same as elsewhere in the USA. Nonetheless, Roseto had unmasked the science behind social engagement and good health and its converse – social isolation, loneliness and early death.

—◆—

Macaques, like Jojo, and the rhesus monkey afford a unique opportunity to study such friendships and relationships. The monkey genome shares 93 per cent sequence identity with the human genome and numerous aspects of their anatomy, physiology, neurology, endocrinology and immunology directly parallel those of humans. Macaques have a lifespan measured in decades and develop, mature and age in similar ways to humans. The ageing process of the macaques looks very similar to our own, including greying and thinning of hair, redistribution of body fat, loss of skin tone, loss of vigour and loss of muscle tone. With age, they get diseases just like ours, including diabetes, cancer, muscle weakness (sarcopenia), bone loss (osteoporosis) and more. The monkeys also exhibit feeding patterns and sleeping behaviour similar to humans. These shared genetics and behaviours mean research in monkeys can often be translated into human observations and studies; extrapolation of the results from monkey studies to human studies is a valuable asset.

Another key advantage of research into monkeys is that we can control factors or variables that are difficult to control in humans. For example, monkeys can be reared in an environment which provides them all with the same diet and the same habitat. With humans, it is almost impossible to control each aspect of life that might affect heart disease. Where the factor being tested is something the monkeys eat, for example, it's possible to give the foodstuff to one half of the monkey group and not to the other, but keep all the other elements of the monkeys' habitat and diet the same, in what is known as a randomised control trial.

Cayo Santiago is a small, palm-fringed 38-acre island haven off the coast of Puerto Rico. It is home to a thriving macaque research station and up to 1,000 free-ranging rhesus macaques. The individuals in the colony are the direct descendants of 409 monkeys who were brought to the island in 1938. The colony is run and maintained by the Caribbean Primate Research Center and the University of Puerto Rico.

It is akin to a school playground. Among the sociable macaques, cliques, best-friend pairs and social climbers are all much in evidence, giving scientists a close look at the primate origins of our drive to affiliate and befriend. After 70 years of research at the field site, the monkeys are well habituated to human experimenters. Due to absence of starvation and predation, this is a perfect system to study social relationships and friendships among our evolutionary close cousins. Just as studies on humans have shown, longevity among the macaques is linked to having strong social connections, including spending time together and time grooming each other's fur, as was the case for Jojo and the kitten.

The macaques of Cayo Santiago provide the perfect opportunity to study the impact of individual friendship and relationships on the ageing process and to understand when these effects start and how long they take to make a difference. In adult female macaques, close female relatives are a proxy for friendships. The number of close female relatives or friends changes depending on an individual's age and this is driven by the need for protection. Females at the peak of reproductive activity (prime-aged) have the most friends and have better survival compared with prime-aged females with fewer friends. These friends provide protection. Older females, however, have experience in navigating the social landscape and so are less frequent targets of aggression, which precludes the need for as many friends. From this, we can see that social support promotes survival and strategies learned across the life course are important to animals in their old age as they consequently require less social support for 'protection'. Social relationships are not just important in macaques but rather are associated with longer lifespan in many other affable species such as baboons, dolphins and rats, implying a common evolutionary basis across species for friendship.

What about social relationship across the life course in us? Although the overwhelming majority of studies to date have focused on links between sociality and longevity in older people, the science has honed in on determining when these links start to emerge and how long they last. In humans, in contrast to macaques, social network size is important for physical health in both young and old adults. We use the 'protection' that friendship affords us both in early life and in later life.

Lisa Berkman, a friend and well-known social epidemiologist in Yale, published some of the earliest studies that detail why social interactions matter and what types of social networks affect our health and even impact when we die. The Yale group used household information from 2,229 men and 2,496 women aged 30 to 69 years who had completed a detailed questionnaire of lifestyle and social contacts, and for whom nine years of follow-up data, including data on the time and cause of death, was available. Overall, 10 per cent of men and 6 per cent of women who had filled in the questionnaire had died, representing 2.2 per cent of men aged 30–39 up to 28 per cent men aged 60–69 years. Four sources of social contact or social 'ties' were examined: marriage, contact with close friends and relatives, church membership and other clubs or group memberships. With few exceptions, people with each type of social tie had significantly lower death rates than people lacking such social ties. Since these studies, many longitudinal studies have reinforced the impact of social ties on death rates.

So why does the strength of social contacts and social engagement affect our mortality? Some explanations include more stress, higher levels of stress-related hormones, more heart disease and more inflammation if we lack strong social ties. Supporting these explanations, Harvard researchers, in a more recent large study of human social networks, found that greater friendship and familial ties independently predicted lower concentrations of fibrinogen, which is a clotting factor in blood and a cause of blood clots and heart attacks, and indicates inflammation. The strength of the association between fibrinogen and social isolation was remarkable. The effect was the same as for smoking, a well-recognised major risk factor for blood clots and heart attacks.

The answer for an association also lies with stress hormones. Biologist Lauren Brent reports that among the Cayo macaques, those with the weakest social networks have the highest levels of stress hormones. High stress hormones trigger a cascade of physiological responses which, if repetitive, lead to heart and brain diseases and early death, further explaining why friendship buffers against disease.

In another experiment, John Capitanio, a psychologist at the University of California, took biopsies of lymph node tissues from monkeys who had been separated from their social group and compared these with biopsies from monkeys within a social group. Lymph nodes are the engine for inflammatory and immune responses. The biopsies showed high activity in inflammatory genes and low activity in genes that protect against viruses. In other words, being friendless had switched on genes known to increase inflammation, the background for a number of age-related diseases. Thus, background inflammation and higher susceptibility to infections is another reason for the association between friendships, disease and mortality. These primate observations align with human observations in Roseto and other social network research.

Reporting from southern Kenya, Lydia Denworth, a science journalist, described the human-like social gestures she observed among a troop of other gregarious mammals, baboons, who spend much of their day embracing, grooming each other and playing with one another's babies. She related the tale of a Botswanan baboon named Sylvia, who scientists dubbed the 'Queen of Mean' because she 'cut a swath through the group, scattering subordinates, biting or whacking animals that failed to move out of her way'. Sylvia's best friend was her daughter who was unfortunately killed by a lion. Sylvia's approach softened after her daughter's death. Bereft of her closest companion, she began offering to groom the peers she'd once scorned, much as a former school bully might try to befriend classmates she'd bullied. This story illustrates how friendship is hard-wired; it is not a choice or a luxury but rather a necessity that is critical to our ability to succeed and thrive. It evolved due to a direct protective bearing on our mental and physical health. Given that Sylvia was hard-wired for friendship, she needed to develop new friends once her daughter was no longer.

Studies also puncture the stereotype that female friendships thrive on endless chats and male ones on side-by-side activity. When pairs of male friends were instructed to ask each other deep questions about their dreams, values and relationships, the men reported being more satisfied with their friendships afterwards.

They concluded that, contrary to popular belief, many male friendships also require depth which may not always be apparent at first sight.

—◆—

Friendship has profound genetic origins. Our closest friends, the people we consider 'kindred spirits', resemble us on a biological level. We share more DNA with friends than we do with other people. A Californian study showed that we share 0.1 per cent more DNA with friends than the average stranger. This may not sound like a lot but it is – it is equivalent to a level of genetic similarity expected among fourth cousins. Most people don't even know who their fourth cousins are, yet somehow, among a myriad of possibilities, we manage to select as friends the people who resemble our kin and thereby we associate with people who are similar to ourselves.

In another series, researchers who studied 5,000 pairs of adolescent friends ran a number of genetic comparisons, seeking to learn more about pairs of friends and schoolmates. Overall, friends were more genetically similar than random pairs of people and about two-thirds as similar as the average married couple. As well as friends being genetically similar, we are also more similar genetically to our spouses. This makes sense – humans naturally gravitate towards people with whom they have something in common! Even our spouses!

Genes drive both friend choice and loneliness. As a clinician, loneliness is one of the most challenging and sad situations to deal with. It is unfortunately a growing epidemic for all ages but most so for older age groups. Vivek H Murthy, the 19th US surgeon general, has graphically detailed the toxic and escalating state that is loneliness. He has provided recommendations for the individual and for society to treat loneliness, squarely placing it as a public health concern and a root cause and contributor to many of the epidemics sweeping the world today, including alcoholism, drug addiction, obesity, violence, depression and anxiety. As with the macaques, loneliness is toxic to human health because of our innate desire to connect. We have evolved to participate in community, to forge lasting bonds with others, to help one another and to share life experiences. We are, simply, better together.

There are a number of key strategies that will help with loneliness. Most are self-evident but, nonetheless, I have described them here. Spend time each day with those you love. Focus on each other. Forget about multitasking and give another person the gift of your full attention, making eye contact and genuinely listening. And it might sound counterintuitive at first, but embrace solitude because the first step towards building stronger connections with others is to build a stronger connection with yourself. Meditation, prayer, art, music and time spent outdoors can all be sources of solitary comfort and joy. Help and be helped. Checking on a neighbour, seeking advice, even just offering a smile to a stranger six feet away all can make us happier and help with loneliness.

Unfortunately, some of the interventions imposed for prevention against the coronavirus such as 'self-isolation' and 'social distancing' have exaggerated loneliness for many. The long-term consequences and implications of these global public health approaches can only be imagined and we should be putting public health strategies in place to mitigate the likely inevitable consequences.

In April 2018, the UK government appointed Tracey Crouch as the world's first minister for loneliness, a role created by Prime Minister Theresa May earlier that year. 'For far too many people, loneliness is the sad reality of modern life,' May said when announcing the new position. The appointment was created in response to a commissioned report that found that more than 9 million people in Britain – around 14 per cent of the population – often or always feel lonely. Loneliness is estimated to cost UK employers up to £3.5 billion annually. From my own research, a quarter of Irish adults feel lonely some of the time and 5 per cent often feel lonely. However, living alone doubles the likelihood of experiencing loneliness. Men who live alone are lonelier than women who live alone. Loneliness increases with age and lonely people are also more likely to suffer from depression. Contrary to our expectations, I found no difference in the likelihood of being lonely for people living in rural Ireland compared with city dwelling.

Probably one of the most striking cultural experiences of loneliness is Japan. Here, lonely deaths among older people have a name, *kodokushi*. This refers to someone dying alone and remaining undiscovered for a long period of time. The first instance that became national news in Japan was in 2000, when the corpse of a 69-year-old man was discovered three years after his death. Monthly rent and utilities had been withdrawn automatically from his bank account and only after savings were depleted was his skeleton discovered at his home. The body had been consumed by maggots and beetles. In 2008, there were more than 2,200 reported lonely deaths in Tokyo. Similar numbers were reported in 2011. One private removals company in Osaka reported that 20 per cent of the their jobs involved removing the belongings of people who had died lonely deaths. Approximately 4.5 per cent of funerals in 2006 involved instances of *kodokushi*.

Kodokushi mostly affects men who are 50 or older. Several reasons for the rise in this phenomenon have been proposed. Social isolation is increasing as older Japanese people are living alone more and more, rather than in multi-generational housing. They lack contact with family and neighbours, and are thus more likely to die alone and remain undiscovered. Japan has the highest proportion of long-lived people in the world. One hopes that this terrible epidemic of loneliness and in particular *kodokushi* is not replicated in other countries where there is also a growing older demographic. Social isolation is commonly coupled with economic hardship. Many incidents of *kodokushi* involve people who were receiving welfare or had few financial resources. The Japanese trait of uncomplaining endurance, or *gaman*, discourages people in need from seeking help. Victims of *kodokushi* have been described as 'slipping through the cracks' between governmental and familial support. Future policy should focus on these high-risk indicators.

Loneliness is not the preserve of old age – it spans all ages. In a recent US survey of over 20,000 people aged 18 upwards, loneliness was reported in all age groups. Social support and meaningful daily interactions had the strongest link with less loneliness, as had good family connections, good physical and mental health, friendships and being in a couple. Social anxiety was most strongly associated with loneliness, followed by social media overuse and daily use of text-based social media.

As you might expect, modern-day change in family structures is strongly implicated in loneliness. Household size is shrinking and there are now more single-person households in Europe than any other household type, which coincides with greater awareness of loneliness as an issue for all ages. How much effort we invest into relationships affects the level of support we receive from them and the long-term benefits. This is the case for all age groups and the benefits persist throughout our lifespan. Close relationships include both family members and friends. But is there a difference in the benefits to health and well-being from each? Should I invest more in friends or in family?

When we speak of family members we mostly include siblings, children, parents and spouses. Harmonious family relationships have a long history of endowing positive effects on people, whether the relationships are with spouses or other immediate family. Friendships enrich health and well-being. William Chopik, a psychologist at Michigan State University, conducted two large-scale analyses to understand the relative contributions of friends and family to good health and happiness across the lifespan, including in later life.

The first study examined over 271,000 people born between 1900 and 1999, and aged from 15 to 99 years, from 97 different countries. Participants were asked how important family and friends were in their lives. They were also asked to rate their own health and happiness. In relation to well-being, participants were asked, 'All things considered, how satisfied are you with your life as a whole these days?' The study was then repeated in a US cohort of adults aged 50 and older, average age 67 years, for whom long-term follow-up data on chronic health conditions, such as high blood pressure, diabetes, cancer, lung disease, heart attacks, angina, heart failure, emotional, nervous, or psychiatric problems, arthritis or rheumatism and stroke, were available to determine whether relationship quality had an enduring impact on long-term health into older ages.

Questions about the quality of relationships included, 'How much do they [close friends/family] really understand the way you feel about things?' and 'How much do they let you down when you are counting on them?' For both studies, spousal support, child support, and friendship support were all associated with subjective well-being and happiness. This was the case at all ages and persisted through to latter years. However, if relationships were strained, chronic disease was more likely. In fact, strain from family and friendships was a principal predictor of likelihood to develop chronic illnesses over time. The findings align well with other research on the overall and lasting benefits of close relationships and the importance of relationship quality rather than relationship numbers.

So, when friends and family are the source of strain, people experience more chronic illnesses; when friends and family are the source of support, people are healthier. Although social networks tend to decrease in size as we mature, we shift more attention and resources towards maintaining existing relationships to maximise our well-being. Thus, as we invest more in relationships over time, we will most likely accumulate the benefits conferred by the relationships, resulting in better health and well-being into older adulthood.

Friendships play a big part in health and well-being in late life because our interactions with friends stem from choice and we are most likely to maintain the friendships that we enjoy most. On days when we positively interact with friends, we report greater happiness and more positive mood. Friendships are more closely tied to well-being because friends often engage in leisure activities together, in limited doses that involve a degree of spontaneity. In contrast, selectively removing family relationships which are stressful or unpleasant is considerably harder than removing strained friendships, which explains why friendships have a stronger impact on happiness than some family relationships.

Strained family relationships impact negatively on health. Family relationships, while enjoyable for many people, can also include more serious and sometimes negative and monotonous interactions. It is, therefore, well worth investing in close friendships for long-term pleasure and better health, happiness and well-being, and sometimes to help you buffer the negative effects of strained family relationships! We should make a conscious effort to put time and attention into building quality relationships. We quite literally cannot afford not to. We should also bear this science in mind if dealing with future pandemics.

What about marriage, health and happiness – till death do us part? Historically, large studies show that, on average, married people report greater happiness later in life than unmarried people. Separated and divorced people are least happy, while the never-married and widowed fall somewhere in between. The positive effects of marriage on happiness are reported by both women and men. But are married people happier because they were happier to begin with? While studies do show that happier people are more likely to get – and stay – married, this does not fully explain the relationship. Happy people who get married still end up happier than happy people who don't. The relationship between marriage and happiness is, like most things in psychological science, bi-directional. In other words, it's what you do to foster happiness as an individual and a spouse that makes a difference, not marriage all by itself. Marriage doesn't necessarily make us happy; rather, happy marriages make us happy. Apologies for stating the obvious but it's what the studies show!

Indeed, when studies measure it, marital satisfaction is a much stronger predictor of happiness than just being married and, unsurprisingly, being in a toxic relationship is decidedly bad for happiness. Single people who elect to never marry but have strong social support through other means are certainly happy and happiness increases when low-quality marriages dissolve. This is true for both men and women. People who stay in relationships that turn sour in order to preserve the ideal – for the sake of appearances, for kids or for basic sustenance – may be married but it hurts both their happiness and their health. Altogether, decades of research on human development, psychology, neuroscience and medicine irrefutably converge on this conclusion: being in a long-term, committed relationship that offers reliable support, opportunities to be supportive and a social context for meaningful shared experiences over time is definitely good for well-being.

Friendship requires risk and commitment but the lifetime rewards that accompany inner-circle ties make the risks and time demands worthwhile. In one decades-long Harvard University study, people who enjoyed strong social bonds into their eighties were less likely to succumb to late-life cognitive decline and dementia. Researchers at Michigan State University tested which aspects of social relations were most associated with memory or remembering in over 10,000 people whose ages ranged from 50 to 90. Participants were tested at two-yearly intervals for six years. Being married or partnered, having more frequent contact with children and friends and experiencing less strain within relationships were each independently associated with cognitive functions such as better memory and less memory decline over time. So the clear message is that frequent quality relationships are good for the brain.

I do want to stress at this juncture that worry about getting dementia, particularly as people develop memory problems, is something I encounter daily and I understand the need for reassurance. Not all memory problems indicate dementia and most deterioration in memory or remembering is age related, very common and does not progress to dementia. Cognitive functions refer to multiple mental abilities that we use regularly throughout the day, such as learning, thinking, reasoning, problem solving, decision-making and concentration. Loneliness and isolation cause a decline in all of these mental abilities Social engagement, connections with family and friends and participation in activities and organisations protect against poor cognitive functioning and dementia.

—◆—

What is the biological explanation for the relationship between friends and the brain? In 2019, colleagues at University College,

London conducted a large review of the published literature exploring the effect of three lifestyle factors on cognitive function and dementia: social networks, physical leisure and non-physical activities. They then summarised all of the evidence, taking into account the limitations of the studies and their biological plausibility. A beneficial effect on brain function and mental abilities and a protective effect against dementia were evident in all three lifestyle components. All seemed to share common pathways, rather than having specific separate pathways, which converge within the three major theories for why we get dementia: the cognitive reserve hypothesis, the vascular hypothesis and the stress hypothesis. We will dwell briefly on each to better understand why friendship changes brain health and why this matters from early adulthood onwards.

Let's start with an experiment in rats to better explain the cognitive reserve theory. Cognitive reserve implies that we have 'banked brain capacity', which is not always used but can be called upon if needed, just like a lifelong savings account in the bank. Environmentally enriched conditions for the rat are equivalent to conditions in the wild, with plentiful opportunities for physical activity, learning and social interaction – rat utopia. Rat utopia prevents cognitive problems in adult rats by building up the rat's brain's 'savings account' – that is, by storing up cognitive reserve. Conversely, an impoverished environment of loneliness and low activity in rats is linked to impaired brain function. The good news is that this is partly reversible by enriching impoverished environments.

Both human and rat brains have the capacity to form new brain cells, new blood vessels and new communication bridges between brain cells throughout life, all of which constitute brain reserve. Mental stimulation, such as that provided by social contacts, physical exercise and creativity, increases the formation of these structures and therefore of reserve. New brain cell formation and therefore cognitive reserve happens predominantly in three key brain areas: in the hippocampus, which sits on each side of the brain and converts short-term memory to long-term memory; in the olfactory bulb, which is at the front of the brain above the nose and governs our sense of smell and in the cerebral cortex, which is important for concentration, understanding, awareness, thought, memory, language and consciousness. So new brain cell formation and cognitive reserve pretty much covers most of our important brain functions. MRI brain scans confirm that people with higher cognitive reserve, due to mental stimulation from social contacts, tolerate more brain pathology. This means that despite having dementia pathology – i.e., abnormal proteins in brain cells – they do not have signs of dementia and live life apparently with normal brain function because they have more 'reserve capacity' to call on.

Social, mental and physical stimulation through friendships and relationships also act via the vascular system. High blood pressure, high cholesterol, irregularities of the heart – particularly atrial fibrillation in mid-life – are all associated with Alzheimer's dementia in later life. Social engagement and relationships reduce these vascular diseases, which in turn reduce vascular causes of dementia and further explain why social contacts protect the brain.

Relaxation and stress reduction are the third explanation for the link between social friendship and dementia. Active individuals with frequent contacts and opportunities to engage with others are more likely to have positive emotions such as high self-esteem, social competence and good mood, all of which lower stress and stress hormones. A higher susceptibility to stress doubles the risk of dementia by triggering chronically high cortisol levels. If you take nothing else from this publication other than a determination to build friendships, it will make a difference to your biological age and a difference to others whom you befriend.

Chapter 4

Never a Dull Moment –
Laughter and Purpose

L AUGHTER OR A SMILE IS THE SHORTEST DISTANCE BETWEEN
two people. We are wired to be happy and share the happy
experience with others through laughter. Laughter is a social
behaviour; we use it to bond and to communicate. You can actually
tell the strength of a relationship between people from the tone and
type of laughter. But you already know this! The laughter of a child
being tickled, the laugh from someone who feels obliged to respond
to their boss's joke and the laughter between good friends are all
different and convey the type of relationship being shared. We
laugh less frequently as we get older, yet the benefits are retained
for life. Laughter is a simple way to boost many of the age-related
cell pathways, which is why it is particularly important to us as we
get older. As well as being a feel-good action, laughter contributes
to better health by muscular exercise, increased respiration and
blood circulation, improvements in digestion, emotional catharsis
and joy. Healthy children laugh as much as 400 times per day
but older adults tend to laugh only 15 times per day. I reflected

as I wrote this that I couldn't recall having laughed today – it was already 6pm!

Most of the time when we laugh it has little to do with humour and more to do with the social bond. We use laughter and humour to manage situations, to display our willingness to engage, to show those present that we are on the same page. We are also more likely to laugh when there are other people around. Friends spend on average 10 per cent of a conversation laughing and clearly we laugh more if we know and like the people we are with. The most important thing on our horizon is other human beings and how they interact with us and what they think about us. Laughter is therefore key to important social interactions and matters because of the interactive role it plays in enabling bonds with others, which are core to survival and still play meaningful physiological and psychological roles, including key roles in ageing.

So, laughter connects us with others. Just as with smiling and kindness, laughter is contagious; we can 'catch' laughter from someone else and are more likely to catch it if we know them. By elevating mood, laughter reduces stress levels in all parties involved.

Laughter has also been described in a number of animal species. This makes sense of course because it is part of our mammalian evolution and, if you think about it, laughter closely resembles an animal call. This is true of some people's laughs more than others! Great apes produce laughter when they are socially playing. Dogs laugh and, in order to set themselves up for play and laughter, they have a 'play bow' which they perform beforehand. Even rats laugh: mother rats tickle offspring inducing a laughter response from the young. Consistent with the bonding role of tickling, it requires at least two animals or people. Try to tickle yourself – it is not possible because tickling is a social interaction coupled with trust. A stranger cannot come up to you on the street and start to tickle you. The intent of tickling is playful, safe and non-threatening, and the result is laughter.

Thus, humour, laughter, learning, bonding and health are all linked. The benefits of humour and laughter are well detailed through history, as early as the reign of Solomon, 971 to 931 BC, in the Book of Proverbs, where it states: 'A cheerful heart is good medicine, but a crushed spirit dries up the bones,' indicating that even then people understood that a joyful spirit has positive therapeutic effects.

Ancient Greek physicians, as an adjunct to therapy, prescribed a visit to the hall of comedians as an important part of the healing process. Early Native Americans used the impact of humour and laughter in healing, coupling traditional medicine men with the services of clowns. During the 14th century, French surgeon Henri de Mondeville used humour to distract patients from pain endured during surgery – there were no anaesthetics until 1847. Even for amputations, de Mondeville used laughter both during and after surgery to aid recovery. He supported this practice in his book, *Cyrurgia*, stating, 'Let the surgeon take care to regulate the whole regimen of the patient's life for joy and happiness allowing his relatives and special friends to cheer him and by having someone tell him jokes.' The English parson and scholar Robert Burton extended this practice by using humour to treat psychiatric disorders in the 16th century, which he discussed in his book, *The Anatomy of Melancholy*. During the same period, Martin Luther, German priest and founder of the Lutheran religion, used humour to treat psychiatric disorders as a critical component of pastoral counselling. Luther advised individuals with depression not to isolate themselves but to surround themselves with friends who could joke and make them laugh. Laughter has a long and successful history in medicine.

So what happens when we laugh? Laughter is fundamentally a different way of breathing. When we laugh, we are using the intercostals (the muscles between the ribs) to repeatedly blow air out from the lungs without inhaling. Laughter also increases the pressure in the chest by effectively breath-holding and stopping the normal regular rhythmic flow of air in and out. The increase in chest pressure lowers blood flow to the brain which sometimes makes people feel dizzy or pass out. This is why the expression 'I almost passed out laughing' holds true.

I run a specialist clinic for adults who experience blackouts and occasionally encounter patients who have an exaggerated physiological response to laughter, where the heart rate slows to a stop, blood pressure consequently drops and they pass out. One memorable patient only suffered these attacks when her son-in-law told a joke. The jokes were invariably smutty and it had become such a common occurrence that her family brought many videos to clinic to illustrate the frequency and characteristics of the blackouts triggered when she laughed a lot. We attached her to equipment that simultaneously measured blood pressure, heartbeat and brain blood flow. We then asked her son-in-law to tell one of his jokes. She duly burst out laughing and then passed out. Her heart transiently stopped, thereby stopping blood flow to her brain. She required a pacemaker to prevent the blackouts and the good-humoured family brought me subsequent video recordings of the patient in peals of laughter with no untoward consequences. In other words, the pacemaker kicked in when her heart started to slow as she laughed and thereby prevented any pause in heart rate.

Laughter provides a physical release; it creates a 'workout'. A good belly laugh exercises the diaphragm, contracts the abdominal muscles and works out the shoulders, leaving muscles more relaxed afterwards. It even provides a good workout for the immune system and the heart.

Is having 'a good laugh' really beneficial at a chemical level? Yes, because laughter lowers the stress hormones cortisol and adrenaline. Low cortisol stabilises blood sugars and insulin, regulates blood pressure and reduces inflammation. Adrenaline is a fight-or-flight chemical, which increases blood pressure and the intensity with which the heart pumps. It is implicated in irregularities of the heart and heart attacks. Its actions are the opposite to relaxation hormones. So, lowering adrenaline calms the nervous and cardiac systems. Diminishing or blocking the effect of adrenaline by laughter has even been shown to work in patients after heart attacks. A single one-hour episode of daily mirthful laughter lowered the rate of recurrent heart attack by 42 per cent.

Laughter also increases endorphins – chemicals produced naturally by the nervous system to cope with pain or stress, the 'feel-good' chemicals. It raises serotonin and dopamine, both endorphins which play a critical role in sensations of pleasure, motivation, memory and reward. They make us feel calm, poised, confident and relaxed. When serotonin and dopamine levels are low, we are nervous, irritable and stressed. Certain substances, especially cocaine and nicotine, are addictive because they stimulate the dopamine-mediated reward system in the brain. How much better to stimulate these systems through laughter, which has no side effects, only copious benefits.

Endorphins are not just about pain and stress, they also play a role in the immune response and in 'killer' T cells, which help to fight infections. Given that immune function declines with age, boosting endorphins is particularly beneficial in older persons. High stress hormones weaken our immune system, so lowering these hormones is one way that frequent laughter benefits immunity and reduces infections.

Even the anticipation of 'mirthful laughter' is good for us. If you expect to have a laugh, your positive hormonal and chemical system kicks in early – even before the laughter starts. One experiment measured levels when volunteers anticipated that they would be watching a humorous video but before the movie commenced. Levels of good chemicals such as endorphins rose to as high as 87 per cent from the baseline level just with anticipation. The same anticipation of laughter muted stress hormones, cortisol and adrenaline, by as much as 70 per cent. So, next time you start to look for your favourite *Father Ted* episode, you are building up your health stores and resources. What would Dougal have to say about that, Ted?

The World Health Organization (WHO) predicts that depression will soon be the second commonest cause of disability worldwide. In depression, brain neurotransmitters, such as noradrenaline and endorphins (dopamine and serotonin), are low and the mood control circuit of the brain malfunctions. Because laughter alters dopamine and serotonin and enhances endorphins, laughter therapy works for patients with depression – either as a single treatment or complementary to antidepressant drugs. There are a number of websites where you can find out more about laughter therapy and laughter yoga. Surely with this long litany of advantages for laughter, we should endeavour to install as much joy and laughter as possible at all stages of our lives? And yes, the number of times we laugh declines as we get older but the potential for physical and psychological advantage is still there – we just need to make more effort to use it.

———◆———

Closely aligned with the health benefits of laughter is having a sense of purpose. Purpose is a key psychological strength which shares many of the biological benefits we get with laughter. One of the first physicians to detail the value of purpose was a psychiatrist who spent three years as a prisoner in Nazi concentration camps, where he documented how purpose had life-saving features. His name was Victor Frankl and he went on to develop a psychotherapy which is still used today and is based on his observations in the camps.

Frankl published the chronicle of his experiences and observations as a prisoner in the camps in his 1946 book *Man's Search for Meaning*. His approach to stress centres around the role of 'having a purpose' irrespective of our circumstances. You can imagine how challenging it must have been for prisoners to find purpose in their lives in the camps. And yet, this is precisely what Frankl described and posited: that prisoners who adopted this approach 'of having a purpose' were better able to withstand the severe stress and awful circumstances. His psychotherapeutic method involved getting his patients to identify a purpose in life, something to feel positive about, and then immersively imagining that outcome. According to Frankl, the way a prisoner imagined the future affected his longevity: 'He who has a why to live can bear with almost any how.'

Frankl maintained that:

> '. . . everything can be taken from a man but one thing: the last of human freedoms – to choose one's attitude in any given set of circumstances, to choose one's own way. And there were always choices to make. Every day, every hour, offered the opportunity to make a decision, a decision which determined whether you would or would not submit to those powers which threatened to rob you of your very self, your inner freedom; which determined whether or not you want to become the plaything of circumstance, renouncing freedom and dignity to become moulded into the form of the typical inmate.'

Frankl concludes that the meaning of life is found in every moment of living; life never ceases to have meaning, even in suffering and death. In one group therapy session, during a time when the camp's inmates were suffering even more greatly as their rations had been taken away as punishment for trying to protect an anonymous fellow inmate from fatal retribution by authorities, Frankl offered the thought that for everyone in a dire condition there is someone looking down. This might be a friend, family member or even God, who would expect not to be disappointed. He used this approach to encourage inmates and provide purpose to their actions.

Frankl described from his experiences and observations that a prisoner's psychological reactions were not solely the result of the conditions of his life but also of the freedom of choice he always had, even in severe suffering. The inner hold a prisoner had on his spirit self-relied on having a hope in the future; only once that prisoner lost that hope was he doomed. This was some of the earliest and most insightful exploration of the value of having purpose. Frankl continued his research and therapies after liberation from the camp. He died in 1997, aged 92; the book has sold over 16 million copies and has been translated into 50 languages.

Today, we know that having a purpose is central to living a happy and longer life. Sometimes, as we get older, families dissipate and we no longer have employment or much in the way of social engagement, and we lose sight of purpose. Life may seem to be aimless and pointless. Purpose is about reflective activities in which individuals perceive their existence to be meaningful and include the goals for which they live.

It is easy to fall into the trap of believing you have no purpose. If a person feels they don't have a purpose in life, they should try to make one. While some people lose purpose when they retire, others take on new challenges. Most volunteering is done by retired persons. An abundance of data shows that people who engage in volunteering are less depressed and have better quality of life. Volunteers are needed in so many different domains in today's world, there should be plenty of choice. Grandparenting provides purpose in a multiplex of ways, contributing significantly to the workforce by galvanising parents to employment and increasing national and individual economic capacity, with a multitude of consequent advantages for the wider family network. Characteristic of many centenarians is a persistent sense of purpose; this is particularly evident in the Blue Zones, where older people have special names for 'getting up each morning with a clear purpose for the day'. Okinowans call it *ikigai* and Nicoyans call it *plan de vida*.

Activities such as choir membership, gardening or a new academic degree, course or diploma are well-established things that can provide a sense of purpose and positive psychological health benefits. Purpose is also amplified through creativity. Neurological research shows that making art improves not just mood but also cognitive function by making thicker and stronger new connections between brain cells. Art enhances cognitive reserve, in other words – the spare brain capacity we have to draw on when called for, helping the brain to actively compensate for pathology by using more efficient brain networks or alternative brain strategies. Making or even viewing art causes brain changes akin to reshaping, adapting and restructuring the brain. According to Bruce Miller, MD, a behavioural neurologist at the University of California, while brains inevitably age, creative abilities do not deteriorate, thus reinforcing the contribution to 'reserve brain capacity'. Imagination and creativity flourish in later life, helping to realise unique, unlived potentials and consolidate crystalline intelligence – the intelligence we get from learning and from past experiences. People involved in weekly art participation have better physical health, fewer doctor visits and less medication usage coupled with better mental health than others who do not participate in creative activity. The benefits last for at least two years after participation.

Aristotle, ancient Greek philosopher and scientist, is one of the greatest intellectual figures of Western history. His thinking strategies were responsible for producing some of the most profound advances in human reason. Our modern society and education focus more on the discoveries resulting from these strategies than on the mental processes through which discoveries are made. We learn about great ideas and the names of the creative geniuses but we are seldom taught the mental processes or creative thinking techniques used to look at the same things and see something different. Albert Einstein said, 'Creativity is intelligence having fun.' It is certainly the act of having new and imaginative ideas and turning them into reality. It is characterised by the ability to perceive the world in new ways, to find hidden patterns, to make connections between seemingly unrelated phenomena, and to generate solutions. We employ creativity in writing, in sculpture, in painting and other means of performance.

I am director of a newly established institute for successful ageing in Dublin. As well as being a busy clinical and research facility, the institute includes a central hub where patients, families and staff can let their creative juices flow. This physical space in the centre of a busy hospital is a source of joy and pleasure that constantly amazes with the new approaches and ideas that are generated through poetry, song, paintings, music and much more.

—◆—

For some, religion provides purpose. Overall, religious involvement, belief and spirituality are positively associated with a host of favourable psychological factors, such as lower depression and anxiety, better memory, better planning and organising abilities and, in general, a longer life. Our research clearly shows a positive relationship between religious practice, heart disease and death, with lower blood pressure and better immunity in religious Irish adults. While some models emphasise the link between private spiritual practice, such as meditation, and health, many others stress the added role of taking part in organised services which are further enhanced by social and cultural factors.

Religious practice is also a coping mechanism and it is difficult to disentangle the positive effects of social involvement, social engagement and meditation from private coping. Although the association between religion and mental health issues, such as depression and anxiety, is complex, the overall association between religion and mental health is positive. In countries such as Sweden where the state is responsible for important aspects of quality of life, such as health and education, religion is not such a strong predictor of life satisfaction. This suggests that religion is, at least in part, a means for fulfilling certain needs where these are difficult to satisfy in other ways.

Several studies have addressed the link between religion and health during life-threatening illness. For example, in people with heart disease from birth, religious faith was positively associated with better quality of life. In people on dialysis for severe kidney disease and people with heart failure and recovering from heart attacks, it was also found to improve quality of life.

So, in summary, it is evident that laughter and purpose are core to longevity and good health. And, importantly, it is within our capacity to ensure that both are central in our lives and to encourage others to realise their potential.

A Good Night's Sleep

Chapter 5

A Good Night's Sleep

WE SPEND ON AVERAGE 26 AND A HALF YEARS OF OUR LIVES asleep, or at least in bed. For some of us, it's a matter of putting a pillow under our heads and falling asleep. However, many experience problems with sleeping, increasing in frequency in mid-life and later. Poor sleep is an all-too-frequent problem with advancing years.

A common misconception is that our brains are inactive during sleep, whereas the reverse is the case. Rather than sleep being an inactive or a passive state, in which both the body and the brain are 'turned off' to rest and recuperate from the day's waking activities, the brain goes through characteristic patterns of activity throughout each period of sleep. It is even sometimes more active when we're asleep than when we're awake. If we have a poor night's sleep, we are more likely to complain of feeling 'low' or depressed, have trouble concentrating and suffer memory problems the following day. In this chapter, I will explain why this is the case and provide some solutions to improve sleep.

So let's go back to basics and explore why we sleep. We are

hard-wired to sleep each night as a means of restoring body and mind. Two interacting systems – the internal biological 'clock' and external factors such as light and noise largely determine the timing of transitions from wakefulness to sleep. These two factors also explain why, under normal conditions, we typically stay awake during the day and sleep at night.

Up until the 1920s, scientists regarded sleep as an inactive brain state. It was generally accepted that as night fell and sensory inputs from the environment diminished, so too did brain activity. In essence, scientists thought that the brain simply shut down. Once they began to record brain activity, by placing sensors over the scalp and measuring brain wave 'electricity' – an EEG (electroencephalogram) – it became clear that sleep was dynamic. We never switch off and sleep goes through repeated stages throughout the night.

The sleep stages are defined by whether or not the eyes are moving. There are four sleep stages and I'll briefly explain them because we spend a lot of time asleep and should understand what's going on! During the first three phases of sleep, referred to as N1, N2 and N3, we are falling gradually into a deeper sleep – the deepest being at N3, during which the eyes are still, referred to as 'no rapid eye movement' or NREM. The final stage is when we dream and is the stage associated with rapid eye movements – REM sleep. These four sleep stages make up a single sleep cycle, lasting between 60 and 90 minutes. Our bodies automatically advance through each stage in sequence, waking naturally after approximately eight hours (if we are lucky). All four stages of sleep are essential for performing critical maintenance and repair within the body. They each serve different purposes and all have characteristic EEG patterns. So, what are the differences among the four stages of sleep and which ones are the most important?

N1 is the beginning stage of each cycle. It lasts for about ten minutes at a time. It's the lightest stage and the easiest to wake someone from. Next stage is N2. Almost 50 per cent of the total sleep time is spent in N2 sleep but, in this relatively brief period, our body physiology is adjusting downwards in preparation for sliding into the restorative N3 later stage of sleep. Physically, during N2, heart rate, breathing and other body functions slow down and our body temperature and blood pressure drops. Sleepers in the N2 stage are harder to wake up than those in N1. N3 is also called deep sleep, or delta sleep, due to the brain's production of long, slow waves called delta waves. We enjoy complete unconsciousness during this phase of sleep and are almost entirely oblivious to external stimuli, including light, sound and movements. Waking is difficult and should we wake, we feel very disoriented (a condition sometimes known as being 'sleep drunk'). It is during this phase that common sleep disorders may occur.

Deep sleep is the most physiologically profound stage of sleep. When we enter this stage, our body releases human growth hormone, a powerful substance that plays a vital role in the repair of both body and brain cells. Built-up waste products are flushed away, tissues are repaired and regrown, bones and muscles are built – especially in growing children – and the immune system is strengthened. Deep sleep is considered to be the most refreshing portion of the entire sleep cycle. It effectively erases the accumulated need for sleep that builds over a normal day of wakefulness and plays a major role in helping clear the brain for new learning the following day. Our longest periods of rejuvenating deep sleep occur in the first two sleep cycles. With each successive cycle, N3 sleep shortens and is replaced by N2 and REM stage sleep. The amount of deep sleep that we get shortens as we age. Stages N1 through to N3 are longest in young children and then gradually decrease with age.

The fourth sleep stage is referred to as REM sleep – when rapid eye movements occur beneath the closed eyelids, the body is paralysed, heartbeat and breathing increase, and dreaming occurs. Because muscles in the arms and legs are temporarily paralysed during REM sleep, we are prevented from 'acting out' our dreams. Although we sometimes awaken convinced that we have been 'dreaming all night', in fact, this is the only stage in which we dream. Among the most important effects of REM sleep are stimulation of learning, processing of the day's experiences and thoughts, and consolidation of memory into long-term storage. Experiencing sufficient REM sleep is essential for normal functioning. The symptoms of insufficient REM sleep include mental problems, such as impaired memory, hallucinations, mood swings and inability to concentrate. Physical problems include lowered core body temperatures, impaired immune systems and, in extreme cases, death.

It is worth spending a bit of time on sleep disorders because over two thirds of us have experienced one or more of these. In my experience, patients and their relatives are often concerned by sleep problems, yet the majority are nothing to worry about and will not cause future problems. Most occur because the brain is overactive during sleep and almost all become more common with age.

I have had a number of interesting patients over the years who have experienced sleep disorders. For example, Peter (not his real name), who had slept perfectly well up to age 73 when he started to get night-time cravings. He would suddenly sit up in bed, go downstairs, fill a plate with food from the fridge, eat it all and return to bed. He had no memory of the event the next morning. This went on for over a year and happened almost weekly. He and his wife thought nothing of these night-time cravings. However, one night, the night 'cravings' changed radically. His wife was woken by the sight of Peter trying to eat the pages from his bedside book. When she tried to stop him, he lashed out at her. The following morning, despite no memory of the event, her black eye served as testimony to the story.

Peter came to me because of a blood pressure problem and, during the consultation, his wife mentioned the sleep behaviours. We performed a sleep study, which included EEG during sleep, and from the results of this coupled with his story, we diagnosed REM sleep disorder. Remember that we are temporarily paralysed during REM sleep? This disorder is caused by a malfunction in the part of the brain that normally suppresses muscle activity while people are dreaming, so that they are not paralysed and can act out their dreams. Peter's brain left him free to move during REM phase sleep, allowing him to go into the kitchen while fully asleep. REM sleep disorder is more common with age and present in 10 per cent of people over 70. It can be treated by drugs which refashion abnormal brain waves. Once treated there were no more night cravings for Peter.

A very well-known disorder is 'sleepwalking', when a person appears awake, their eyes open, but they are actually asleep. Again, this is quite common: one in ten of us will experience sleepwalking at some stage in our lifetime, with some people sleepwalking regularly throughout their life. Fortunately, it is not associated with any significant underlying health problems, although the act itself risks accidents.

Bedwetting is another sleep disorder and is common in children but can occasionally persist into adult life, and in some cases gets worse with age. Others gain control after childhood but relapse later in life. As we age, the number of times we pass urine at night increases. For men, this is a particularly common issue – the prostate gland often enlarges with age and presses against the bladder, irritating it and resulting in a more frequent need to pass urine. It can be managed 'conservatively' by ceasing intake of fluids after 4pm. There are also effective medications to control the bladder irritation.

'Night terrors' occur in 10 per cent of children and are most often seen between the ages of 3 and 7. Most of us grow out of them but they persist in 2 per cent. I well remember a very sad and spectacular case attributed to night terrors some years ago in the UK. A retired miner, a 'decent and devoted' husband, strangled his wife of over 40 years during a night terror. He had suffered with persistent night terrors since childhood. He was in the throes of a nightmare about a 'boy racer' who had broken into their camper van where they were sleeping and he dreamed that he was fighting him off. While still asleep, during the 'fight', he strangled his wife. He awoke and rang 999, weeping that he had killed his wife. The miner was acquitted because of his known long history of night terrors and his clear devastation at what had happened. The Crown Prosecution Service accepted he had not been in control of his actions and that he was not a danger to anyone else.

This was of course an exceptionally rare and tragic case. In the majority of cases of night terrors, people are jolted from slumber, terrified, confused and unable to communicate. They may thrash about or get out of bed during episodes. Night terrors occur during deep sleep, don't require specific treatments and generally are not indicative of any worrying underlying problem.

'Sleep paralysis' is a frequent adult sleep disorder where we experience a sensation of being unable to move the body or limbs when either falling asleep or waking up. Almost two thirds of us experience sleep paralysis at some time. Unless it is frequent and troublesome it does not require treatment and will not harm. Some of my patients who experience this believe they have had a mini stroke or are at risk of a stroke – this is very understandable given the symptoms but is not so.

I have personally experienced sleep hallucinations and they are not pleasant. In one of my posts as a young junior doctor, I was on a 'one in two' rota, which meant working every day 8am to 6pm including 24-hour call on alternate days. The hospital was particularly busy and I was in a perpetual state of fatigue. When sleeping, I would wake up convinced that I heard my pager going off and ring the switchboard to find that no one had bleeped me – this was an auditory hallucination and I had heard a non-existent emergency bleep. Once I moved to a more humane rota and got more sleep, the hallucinations disappeared. I am not alone in my experience of sleep hallucinations: one in four of us experience hallucinations associated with stress or fatigue and they occur at all ages.

Sometimes, if hallucinations are not caused by stress or fatigue but are repetitive and frightening, with a person believing that they see, hear, touch or sense something that isn't there, then a form of epilepsy known as narcolepsy, may be the cause. This should be investigated because it can be treated.

Although it is common for people in many Western societies to sleep in a single consolidated block of about eight hours during the night, this is by no means the only sleep pattern. In fact, following this schedule and foregoing an afternoon nap would seem highly abnormal to many people around the world. In cultures with roots in hot regions, afternoon napping is commonplace and built into daily routines.

Afternoon naptime typically coincides with a brief lag in the body's internal alerting signal. This signal, which increases throughout the day to offset the body's drive to sleep, wanes slightly in mid-afternoon, giving sleep an advantage and edge over the wakefulness drive. Napping typically happens during the warmest period of the day and generally follows a large midday meal, which explains why afternoon sleepiness is so often associated with warm afternoon sun and heavy lunches. This is the worst time of day to have to deliver a lecture, particularly to an older audience, for whom napping after the midday meal is even more likely.

For some people, 10-minute 'power naps' are effective. Others find 20 minutes sufficient. Ninety-minute naps, provided that the NREM-REM cycle is complete, are also good to re-vitalise or re-boot. However, it depends on one's individual sleep behaviour and, with time, we learn what period of napping best suits us. If a person has insomnia, then afternoon napping may confuse the body clock and make the insomnia worse. If you nap, try to do so before 3pm. As people get older, they have more fragmented sleep and this is not uncommonly associated with napping by day. For some, this is necessary to preserve function and regenerate energy. For others, daytime napping exacerbates nocturnal sleep problems. It's best to work out one's own napping preference and stick with that, albeit in the knowledge that needs and patterns may change with age.

We can use sleep to enhance learning. Periods of sleep following learning consistently enhance our ability to retain the content of the material. And of course, the converse is true, which is not surprising considering the role of sleep in memory consolidation: insufficient sleep reduces cognitive ability, including concentration, memory and learning.

———◆———

When it comes to sleep and anxiety disorders, William Shakespeare's Macbeth had it right when he referred to sleep as the 'balm of hurt minds'. Research from the University of California, Berkeley has shown that while a full night of slumber stabilises emotions, a sleepless night triggers anxiety levels to rise by as much as 30 per cent. Some 40 million American adults have anxiety disorders and this figure continues to rise. The type of sleep most apt to calm and reset an anxious brain is deep NREM sleep. Sufficient deep sleep decreases anxiety overnight by reorganising connections in the brain, restoring activity in areas that regulate emotions and lowering heart rate and blood pressure. Thus, sleep is a natural, non-pharmaceutical remedy for anxiety. The amount and quality of sleep from one night to the next predicts how anxious we feel the next day. Even subtle nightly changes in sleep affect anxiety levels. So what can stop us from getting enough NREM deep sleep?

Doing vigorous exercise, like fast walking, late at night alerts the sympathetic system and releases stimulating hormones and neurotransmitters that make it harder for our minds and bodies to switch to deep sleep mode. So it's best to take exercise earlier in the day rather than before bedtime. Some people find that eating a late evening meal leads to a troubled night's sleep, while others find it helps. So it's best to work out which suits you. Certainly the content of food and drink before sleep matters, and the ability to tolerate late eating declines as we get older. Aged cheese, bolognaise sauce, bacon and other cured meats, including sausage, pastrami, corned beef and ham, contain hefty doses of tyramine, an amino acid that triggers brain alertness. Some Italian wines and several beers also have a high tyramine content. Tyramine stimulates the production of noradrenaline, a neurotransmitter that is part of the sympathetic nervous system's fight-or-flight response, meaning that we feel alert and awake – ready to fight or flee! Chocolate and coffee contain caffeine, which is also a stimulant. High carbohydrate meals can disturb sleep, as can acidic foods and spicy foods. Foods that are high in fibre, such as broccoli, cauliflower and carrots, can be challenging for digestion at night and are best taken earlier in the day. The traditional 'night cap' is also problematic because, although we will get to sleep more quickly after alcohol, sleep cycles and length of both NREM and REM are disturbed. As the alcohol is metabolised, awakenings become more frequent, which is why alcohol disrupts sleep. The response to these stimuli is genetically programmed – in other words, some people have no problem taking tyramine- or caffeine-rich foods at night. Later on, I will discuss foods that help NREM sleep.

There has been an explosion of research into new technologies to promote NREM sleep. Using sound stimulation, such as listening to pink or white noise, can enhance deep sleep and lead to better memory function the next day. White noise contains all of the frequencies audible to the human ear. Pink noise is white noise with fewer higher frequencies. It increases the intensity and slows down the speed of NREM waves, thus allowing more time to clear toxins, improving the retention of learning and memory and reducing anxiety. It doesn't work for everyone but some report good results with it.

Other promising and exciting new technologies have not been fully validated although some are commercially available. A popular new technology is a band worn around the head with sensors to detect and track slow brain waves. A stimulus is applied by the band to the slow waves, which causes them to lengthen and slow even more, all of which may deepen NREM sleep.

The optimum sleep period for adults irrespective of age is seven to nine hours. Our TILDA research has shown that getting less than seven and more than nine hours of sleep after the age of 50 are both linked to future problems with mental abilities, such as memory, concentration and learning. During slow wave sleep (NREM), the spaces between brain cells fill with additional fluid, called the cerebrospinal fluid, which bathes the brain and the spine. This washes away toxins that have accumulated during the day, including those implicated in dementia – beta amyloid and tau protein. It is important that these toxins and waste products are regularly cleared by the cerebrospinal fluid otherwise they will accumulate and block transmission of signals between brain cells. One very elegant experiment showed that missing sleep for even one night was associated with higher tau levels compared with those after one good night's sleep in healthy middle-aged men. Given that missing one night of sleep results in an increase in the level of tau, it is likely that over time, repeated sleep disruption will have long-term detrimental effects on the brain and mental abilities. Therefore, insomnia in mid-life should be treated with the same gravity as high blood pressure and diabetes – all of which are risks for poor brain health in later life.

—◆—

As young medical students, we took advantage of every opportunity to attend the different faculty balls each year. It didn't matter which discipline held the ball – a ball was a ball. So we went to the Arts Ball, the Ag Ball (Agriculture), the Commerce Ball, the Law Ball and, of course, the Med Ball. The one beauty tip at the time that I clearly recall, which was a mantra among the dedicated female 'ballers', was to be sure to get a good night's sleep before the ball so that your skin looked fresh without dark rings around the eyes – our 'beauty sleep'.

We now have a biological explanation for this axiom. Researchers from the University of Manchester have discovered that 'looking as fresh as a daisy' after a good night's sleep has fundamental biological roots. Collagen is one of the most abundant proteins in the body, responsible for a third of our body architecture. We can think of collagen as the body's scaffolding. It supports the skin, tendons, bones and cartilage. Collagen provides the body with structure, ensuring integrity, elasticity and strength. It is closely aligned with sleep and with age. There are two types of collagen – a thick fixed type and a very thin 'sacrificial' collagen. A good analogy for the two types of collagen are the bricks in the walls of a room which are the permanent part (thick, fixed type), and the paint on the walls, which are the non permanent part (thin collagen fibrils). The collagen fibrils need to be replenished regularly because during the day, this thin collagen takes the brunt of wear and tear and is replenished during sleep. This process is governed by genes but may not work as efficiently as we get older. We look so much better after a good night's sleep because we have replenished the sacrificial collagen necessary for skin integrity, including around the eyes where skin is thin and dark rims easily form.

———◆———

How often have you been asked by your doctor if you snore? Hardly ever? Or even never? Yet snoring can be an early sign of underlying health problems. Clearly, if you have a partner, you will be made aware of your own regular snoring. If you sleep alone then waking with a dry mouth may be a clue. Heavy snoring is linked to a disorder called sleep apnoea, characterised by pauses in breathing during sleep. If the pauses last for ten seconds (long enough for one or more breaths to be missed) and occur repeatedly, then oxygen levels drop. The reduction of oxygen to the heart can lead to heart attacks, strokes and decline in memory and concentration. When oxygen drops, stress hormones surge. Those hormones contribute to high blood pressure, which most people with sleep apnoea experience. Sleep apnoea is present in 3 per cent of 20–44-year-olds, 11 per cent of 45–64-year-olds, rising to 20 per cent of people over 60. It is diagnosed by an overnight sleep test called a polysomnogram, where wires on the head and chest track brain waves as well as heart and breathing patterns.

Anyone who snores loudly, wakes without feeling rested or has high blood pressure, diabetes or is overweight is more prone to sleep apnoea. It is important that this is addressed as the risk of adverse health consequences declines dramatically with treatment. A very effective treatment, one that helps 90 per cent of those who adhere regularly to the method, is use of a face mask that prevents obstruction at the back of the throat by changing the pressure in the mouth and throat. The mask system is referred to as a continuous positive airway pressure (CPAP) device. Simple snoring that is not very loud and without any other symptoms can usually be handled by turning on one's side.

———◆———

Sleep matters when it comes to our susceptibility to infection and to fighting infections when we do get them. During sleep, the immune system releases proteins called cytokines, whose main role is to target infections. Some cytokines also help to promote sleep. Sleep deprivation reduces both the production and the release of protective cytokines, causing a double whammy if you skimp on shut-eye.

Moreover, the contribution of sleep to fighting infection isn't just down to cytokines. Sound sleep also improves the action of immune T cells on fighting infection through a sticky tactic! 'Killer' T cells attack viruses such as influenza, HIV, herpes and Covid-19, by getting in direct contact with them, sticking to them and destroying them. The 'sticky' substance known as integrin is vital to allow the T cells to do this effectively but stress hormones such as adrenaline and noradrenaline block the stickiness of integrins. Because the levels of these stress hormones are low during sleep, there is a higher level of integrins in the body and they are stickier and thus better able to assist T cells in fighting off infection. Good sleepers are less likely to get winter colds and flus and are better able to fight infections should they occur. Chronic poor sleepers get more colds and flus and even have a poorer response to vaccinations. So, in summary, there are lots of 'immune system' reasons to work on better sleep.

◆

To fully understand why sleep matters for ageing we must focus briefly on circadian rhythms – our very own internal clock. Circadian rhythm has shot to the forefront of medical research in recent years. It is present in all living organisms and plays a major role in the changes that speed up ageing. Every cell has an internal clock that drives its circadian rhythm, which is synchronised to all other cells. Circadian rhythms are present to make the most of the cell's abilities, to ensure that energy is not wasted and to give the cell and the body enough of a chance to clear all toxins which would otherwise build up, accelerating ageing and then cell death.

A good example of a plant's circadian rhythm is that of flowering desert plant Mirabilis multiflora, or the Colorado four o'clock plant. During the day, its flowers are tightly shut. After four o'clock, the blooms open for pollination and wilt the next day. For the petals to open, they require water to be transferred from the rest of the plant but given that it is a desert plant, water is scarce. The plant is pollinated by a nocturnal moth and so uses a 'clock' system to open its petals at four o'clock when it is cooler and the moth is circulating. This circadian clock ensures that the plant conserves water as much as possible during the hot day, as well as working efficiently to maximise pollination opportunities at night.

Much like the Colorado four o'clock plant, our cells' clocks work in synchrony – that is, at the same time and same rhythm, through a central control system located in the brain, called the suprachiasmic nucleus, or SCN. This is our master clock and it organises the clocks in every cell of our body for efficiency. It responds to the external cues of light, dark and food, and then orchestrates the clocks in all cells. The SCN helps us to wake up and be alert, tells us when it's time to eat, ensures that our gut is awake and prepared to take in food and tells us when it's time to go to sleep. The SCN is stimulated by light coming from the eye, which is why light and dark control circadian rhythm. All of the measures taken during a visit to your doctor – such as blood pressure, heart rate, temperature, blood levels of lipids, melatonin and cortisol – have circadian rhythms and therefore vary throughout the day. For example, blood pressure is lowest at night when we are asleep, peaks early morning and then settles to a daytime level which sometimes drops a little after a large meal or if we are resting. The SCN and your internal circadian rhythm is responsible for these blood pressure fluctuations. Ageing is closely linked to circadian rhythms and to maintaining a balance between sleep/wake and eating times.

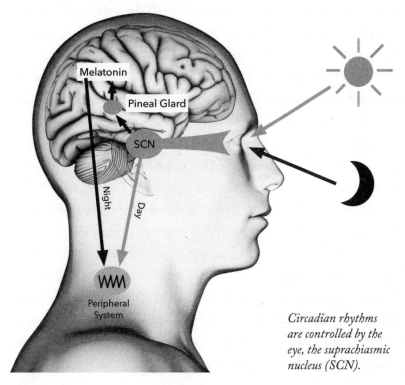

Circadian rhythms are controlled by the eye, the suprachiasmic nucleus (SCN).

The principal timekeeper gene that controls the clocks is the Bmal1 gene. Until 2020, Bmal1 was assumed to be the only time-keeper gene but researchers at the University of Pennsylvania found that skin and liver cells retain 24-hour circadian rhythm even after getting rid of this gene, indicating that while the Bmal1 gene heavily influences circadian rhythm, other genes are also involved. If we could manipulate these genes to be more efficient, it would slow down cell ageing.

A key factor in the link between light–dark stimulation of circadian rhythm, ageing and sleep is melatonin. Melatonin is the hormone which regulates the sleep–wake cycle. Think of it as our body's own sleeping tablet. It is primarily released by the pineal gland in the brain in response to darkness. Melatonin's actions are not confined to sleep regulation – it also has antioxidant properties and beneficial effects on the immune system. In adults, it is mainly produced during the dark phase and the highest blood concentration occurs 4–5 hours after darkness. Light stimulus blocks the production of melatonin and as a consequence, during the daily light period its level is very low. Melatonin production decreases with age. With age, vision also declines and eye diseases such as cataracts become more common. These combine to reduce the intensity of the response of the eye to light and to lower melatonin further and reduce SCN stimulation. Early recognition and treatment of eye problems will help to minimise any negative effects of ageing on the SCN and melatonin and therefore on sleep. This is one reason why regular eye tests are recommended after age 40, which is when age-related eye problems may begin.

With age, we experience a longer delay from sunset to the onset of melatonin rise and the melatonin peak. The relationship between age, declining melatonin production and increasing insomnia led to the 'melatonin replacement' hypothesis. Research shows that replacing the deficiency in this sleep-regulating hormone improves sleep. 'Slow release' melatonin tablets appear to be more efficient than faster acting melatonin. Melatonin at a dose of 2mg for up to 2 years has been approved for the short-term treatment of insomnia in people aged 55 years and over. This is a safe treatment with few if any side effects. Melatonin is also used for the short-term treatment of sleeping problems from jet lag or shift work.

—◆—

There is a deep-rooted connection between fires and human well-being. It's lovely to sit around a 'good fire', which provides not only warmth but also comfort and relaxation, attributed in part to fires emitting yellow light. The ability to start and manage fire allowed the development of cooking and expansion of humans' diets, and was a big part of our evolution as a species. Cooking also played a role in the expansion of our brains. The hearth formed a social focus, helping the development of language. Concrete evidence for the use of flints to start fires comes from as recently as 40,000 years ago but may have occurred as far back as 400,000 years ago. Therefore, until recent history, humans were predominantly exposed to, and their lives and evolution depended on, yellow light (wavelength 570–590nm) and blue light (wavelength 450-495nm) exposure was limited to a few hours in winter. Even the incandescent light bulb, widely used in the 20th century, produced relatively little blue light.

However, over the past few decades, blue light has been used more and more in modern communication technologies, emitted from devices such as televisions, phones and computers. Blue light suppresses melatonin proportional to the light intensity and length of exposure, thereby causing sleep disorders and insomnia. The figure below shows how much blue light affects sleep. The longer the exposure before sleep, the shorter the duration of sleep.

Email checking has the most striking effect, reducing duration of sleep by one hour if exposure goes from none to four hours. It is likely that the negative effects of blue light are exaggerated with age and therefore should be treated with even more caution. Glasses that block blue light in the hours before bedtime improve melatonin levels.

Sleep duration and hours of screen use among 9,846 adolescents

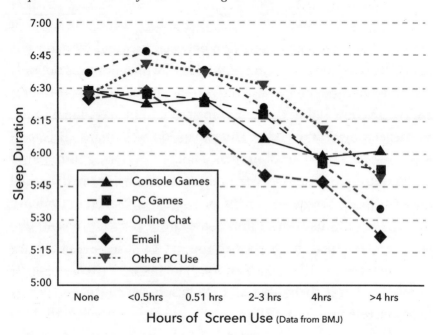

Although the 24-hour circadian clock is strictly regulated by the master clock in the SCN and complemented and assisted by melatonin, we do not all have the same cosy 24-hour relationship with our clock. Some of us are hard-wired to have our own preferred circadian pattern which does not exactly align to light/ dark circadian cycle. This matters because it helps to explain why some of us struggle with early rising when it's light and with hitting the sack at darkness. This alignment to the circadian clock is called our 'chronotype', which is our very own personal circadian 'hard wiring' that governs circadian rhythm. Our chronotype describes our body's natural timeline for the daily basic activities, such as eating and sleeping. The stereotypes of lark or owl represent chronotypes.

Identification of the gene responsible for chronotype, PER3, was the subject of the 2017 Nobel Prize in Physiology or Medicine, which went to three American scientists. Jeffrey C. Hall, Michael Rosbash and Michael W. Young discovered that chronotype is hardwired into sleep schedules, helping to explain why they are difficult to change. This gene is a member of the Period family of genes, responsible for circadian control of how fast we walk, metabolism of sugar, fats and sleep behaviours. It determines whether we are larks or owls. An owl chronotype struggles to function at full capacity in the mornings, whereas a lark is programmed to slow down at night time and is buzzing in the morning. However, although chronotypes are hard-wired, they do change with age.

If we drill down into chronotypes, taking a deeper dive into personalities and characteristics, we find four sub-types of lark and owl: dolphin, lion, bear and wolf. Overall, 10 per cent of people are dolphins, 20 per cent lions, 50 per cent bears and 20 per cent wolves.

Dolphins and lions are early wakers, while the wolf comes to life later in the day and dislikes early morning. Bears' sleep behaviours are between that of early wakers and wolves. Most of us are bears. The first three categories – dolphin, lion and bear – function well within the set timetable that society has imposed on us regarding school and work. Because the wolf's circadian rhythms, and therefore the clocks in all cells, are somewhat out of synchrony with light–dark signals, this group comes to life later in the day and likes to stay up late and work later in the day. Wolves are a minority and society's established timetables do not serve the poor wolf. Consequently, these nocturnal sub-types can be disadvantaged and commonly chronically fatigued or 'socially jetlagged', thus experiencing slow mental processing, day-long hunger, a feeling of exhaustion and apparent 'laziness'.

The 24-hour circadian rhythms of all important physiological parameters, such as blood pressure, cortisol, heart rate, adrenaline, melatonin and temperature, behave differently in the sub-types and are delayed or even flipped in wolves compared with other groups. Because the wolf's hunger and appetite clock is out of sync, wolves are more likely to experience excessive eating and obesity. As a consequence, wolves are more likely to get diabetes, heart disease, strokes and sleep apnoea. Wolves are also more likely to have addictive personalities, encompassing overeating, smoking and excessive alcohol intake. As we age, we morph towards being a dolphin or lion.

Dolphins	• Struggle to fall asleep • Sleep approximately six hours • Wake up unrefreshed • Tired until late evening • May experience anxiety irritability • Highly intelligent • Perfectionist
Lions	• Medium sleep drive • Wake up early • Lots of energy • Little energy at bedtime • Optimistic • Overachievers • Go-getters

	• Health conscious • Eat well • Take exercise • Leaders
Bears	• Deep sleep • Rise with the sun • Strive to be healthy • Team player • Hard-working • Easy to talk to • Good people skills
Wolves	• Wake in a haze • Groggy in the morning • Energy in the evening • Tend to miss breakfast • Come alive after dark • Creative • Pessimistic • Moody • Comfortable alone • Most likely to be addicted of all the chronotypes.

Whereas lions are frequently high-achieving goal-setters and team leaders, wolves tend to be more creative. If you are a wolf and wish to change in order to accommodate society's timetabling, you're not without hope. It is possible to adjust your schedule somewhat to become more of an early bird by gradually changing sleep and food intake by 15 minutes per day until desired sleep/wake times are achieved. But more important, perhaps, is to be aware of one's 'type' and to be on guard against susceptibility to compulsive behaviours and poor lifestyle, and to ensure extra care with diet, physical activity and other habits.

Remember, chronotype is as much about food as sleep! For all chronotypes, restricting eating to within an eight-hour period each day will reduce obesity. Some elegant rat experiments underscore this. If rats with 24-hour access to food are compared with rats who have 8-hour access to the same type and same amount of food, and both finish all the food, the group with 24-hour access become obese but the eight-hour access rats do not. The same applies to humans. A 16-hour fast each night improves sugar tolerance and reduces weight and blood pressure, in addition to stabilising circadian rhythm.

Digestion of food messes with sleep. For a pre-bedtime snack there are a host of sleep-promoting foods that enhance melatonin and neuropeptides such as tryptophan and serotonin. They include almonds, turkey, chamomile tea, kiwi, tart cherry juice, fatty fish (salmon, tuna, trout, mackerel), passionflower tea, white rice, milk, bananas, porridge and cottage cheese. Others such as chamomile contain apigenin, an antioxidant that binds brain receptors to induce sleepiness. Vitamin D and omega oils also improve sleep. In a randomised control trial of 95 men, all components of sleep were significantly better in the group who had omega oil rich Atlantic salmon three times per week compared to the control group who had similar nutritional equivalents but of chicken, pork and beef. In another study of 1,848 people aged 20 to 60 years, high rice consumption before sleep was associated with better sleep than eating bread or noodles.

Whereas problems with sleep become more common as we get older, there is plenty of scope to better understand chronotypes, circumvent poor patterns and modify factors which will help the quality of our sleep and therefore our quality of life.

Chapter 6

Downtime and the Pace of Ageing

OVER THE PAST 30 YEARS, THERE HAS BEEN A SEISMIC CHANGE in our pace of life. Despite all of the electronic conveniences that should theoretically make more time available for activities such as having coffee with a friend, reading a book or just relaxing, it seems we keep getting busier and more stressed. When email and other internet tools were first introduced, they were heralded as the solution to overwork and stress. We were promised utopia: we would work more efficiently and have more leisure time, more time with friends and family, more time to chill, to exercise. We could look forward to shorter working days and longer holidays. Instead, life became even busier and more stressful. Devices are constantly pinging and ringing. As I started to research 'stress', I realised how much I am reliant on devices and how difficult I find it to extract myself from the dependency and take my own advice!

While advancements in technology are amazing, they come at a cost. The persistent beeping, vibrating and flashing of notifications mean that we are constantly distracted and driven to interrupt what we are doing to check our phones. One UK study found that, on average, young adults unlock their phone 85 times a day and use it for a total of five hours each day. This equates to a third of waking hours. Yet awareness of use is underestimated. When asked how often they had used their phones, respondents underestimated by 50 per cent. The consequences are inability to focus attention and consolidate things properly into memory, causing distress.

As evidence for the technology addiction, in a study where young adults were instructed not to use phones, they exhibited withdrawal symptoms, comparable to those seen in someone with a drug addiction. This is backed up by research showing correlations between high smartphone and internet use and poor cognitive skills such as concentration, memory and learning. In one review of 23 published research papers, relationships were evident between smartphone use and depression, anxiety, chronic stress and low self-esteem. Other problems are caused by using the phone right up to bedtime. How familiar is the following: you get into bed intending to go to sleep but decide to check your phone (just for 'a second') to find out something innocuous and then, an hour later, you are still watching something? Couple this failure to disengage and relax with the further negative impact of blue light on circadian rhythms and melatonin and you have a perfect recipe for a poor night's sleep.

Most of the evidence cited for poor outcome from technology use relates to younger adults; the relationship between technology and older adults is complex. In general, internet use by older adults is much more moderate and associated with better mental health and life satisfaction, spurring researchers to encourage use of technologies with age. However, older persons who are not technologically savvy can become disengaged from and marginalised by our rapidly changing digital society. Most of today's services cannot be accessed without the internet, which results in some people feeling further disenfranchised and frustrated.

Few if any of us have not experienced stress at some stage in our lives and older readers will have experience of both acute stress and the chronic accumulated consequences of stress. The heavy duty medical definition of stress is 'the quality of experience, produced through a person–environment transaction that, through either over arousal or under arousal, results in psychological or physiological distress'. This cumbersome definition is hardly necessary – we recognise what we mean by stress, we know how stress feels. Stress is identified either by our own feelings or by objective measures. Biological measures of stress are many and include reactions in the nervous system, hormones, immune or inflammatory systems and metabolic systems. The health consequences of stress are not pleasant and include obesity, diabetes, high blood pressure, a fast heart rate, heart attacks and strokes.

Usually, stress affects a number of systems together, not just a single one. The phenomenon of rapid greying of hair due to acute stress has captured storytellers' imagination like few other afflictions. The likely explanation for overnight or rapid greying is that stress induces loss of pigmented hair with retention of non-pigmented and thus white or grey hair. Rapid greying is often called 'Marie Antoinette syndrome'. The name refers to Queen Marie Antoinette of France (1755–1793), whose hair, allegedly, turned white during the night before her walk to the guillotine at the time of the French Revolution. She was only 38 years old when she died. This is evidence of the acute and extreme stress she experienced on that night. History also records that the hair of the English martyr Sir Thomas More (1478–1535) turned white overnight in the Tower of London before his execution. More modern accounts refer to the hair of survivors of bomb attacks during Second World War turning white. In a published case I came across, an American dermatologist described a 63-year-old male patient who he attended after the man had fallen down the stairs; his hair turned white after the accident, again reflecting the shock and stress for that individual. Senator John McCain was a member of the US Congress from 1983 until his death in office in 2018 and a two-time US presidential candidate. His biographer describes how, as a prisoner of war in Vietnam, McCain's hair rapidly turned white as a result of gruesome torture.

Harvard researchers have shed further light on how stress causes such rapid greying and what this tells us about the wider impact of stress on our bodies and biological systems. Stress activates nerves that are part of the fight-or-flight response – the sympathetic nerves. When scientists explored the impact of different degrees of stress on the hair follicle they found that the sympathetic nerves, which feed each hair follicle, release the chemical noradrenaline. The intensity of noradrenaline release equates to the intensity of the biological stress response and depletes hair pigment in addition to accelerating hair loss. Because the sympathetic system is involved in innervation of almost all organs, the authors concluded that greying is indicative of the pervasive biological effects of stress. We only have a finite amount of hair pigment and once depleted, it cannot be replenished – once grey always grey (visits to the hairdressers excluded). This is why ageing is associated with grey hair. We have evolved to gradually tap into the pigment source over many years and not to empty our stores of pigment over a short period.

The principal author of this Harvard study stated, 'When we started to study this, I expected that stress was bad for the body – but the detrimental impact of stress that we discovered was beyond what I imagined. After just a few days, all of the pigment-generating stem cells were lost. Once they're gone, you can't regenerate pigment anymore. The damage is permanent.' By understanding precisely how stress affects these cells, the authors have laid the groundwork for discovering how stress impacts other tissues and organs in the body – the first critical step towards eventual treatment that can halt or revert the adverse impact of stress and possibly of accelerated ageing.

The good news is that we become less stressed with age. In a large Gallup poll from 140 countries, of respondents aged 15 to 29 years, 64 per cent reported they were stressed, 50 per cent reported they were worried and 32 per cent were angry. In contrast, respondents aged 50 and above were less stressed (44 per cent), less worried (38 per cent) and less angry (16 per cent). And these figures were even lower again in respondents over 70. Another large series from the University of Southern California similarly showed that ratings of daily perceived stressfulness yield a paradox, with high levels from the twenties through to about age 50, followed by a precipitous decline through the mid seventies and beyond. Compared with 50 per cent of younger participants, only 17 per cent of older participants were stressed. Many factors such as less financial pressure, retirement, grown-up families and a more positive outlook contribute but they don't entirely explain the reduction.

This nicely aligns with our research on life satisfaction and happiness, which similarly shows a U-shaped curve – happiness is high in our twenties but drops thereafter, bottoming out in our forties and fifties, then rising again and continuing to increase until our seventies. Life gets better for most of us from age 50 to our mid seventies, after which quality of life slowly and gradually starts to decline, mostly due to physical ill health. However, quality of life scores do not reach the same levels as those seen at age 50 until after our eighties – so we have, on average, 30 good years after age 50 for good quality of life, contrary to popular views.

This is explained by the fact that with age comes more realistic expectations and people become increasingly selective about how, and with whom, they spend time. Older persons are wiser, more likely to live in the present, take each day as it comes, savour the positive, dwell less on the negative, overreact less, set realistic goals and prioritise people and relationships. We become more adept at dealing with stressful challenges as we age and we have more accumulated wisdom, which buffers stress and helps us to cope. The potential for wisdom is at least partly neurobiologically based. In other words, it's hard-wired. Brain-imaging supports a biological explanation for wisdom, showing areas of the brain consistently lighting up when we carry out tasks which involve wisdom, particularly empathy, decision-making and reflection.

Imparting wisdom, through inter-generational sharing, positively influences mental health and well-being and helps to reduce stress in young and old. Teresa Seamen is the senior researcher behind a new and innovative experimental programme of mentoring that illustrates evidence-based benefits for this 'wisdom sharing' experience. It is referred to as Gen2Gen and started in the city of San Jose, California. Designed 'to make great things happen with older adults through volunteering for mentorship, and to lift the aspirations and opportunities of thousands of young people', retirees help to educate under-privileged children and young adults for an average of 15 hours per week. Retirees are paid stipends and engage pro-actively in a strategy and planning programme, and give regular feedback to the formal educators overseeing the scheme. The programme is a huge success, both in terms of enhancing education and destressing the young and the retirees. It's a simple model that's proven to work and will hopefully be replicated in other countries.

—◆—

Believe it or not, we need to turn to inflammation yet again to understand some of the biological markers of stress and how it affects diseases. Both acute and chronic stress trigger an inflammation cascade. This inflammation builds up with time, creating the familiar chronic age-related conditions, such as heart disease, cancers, Alzheimer's – and more wrinkles!

Whereas people in the Blue Zones do experience stress, they have developed buffering techniques against it from which we can learn. What the world's longest-lived people have, that most of us do not, are routines to shed stress so that it does not become chronic. Okinawans take time each day to remember their ancestors; Adventists pray; Ikarians take a nap and Sardinians share wine with friends and family. Life unfolds more slowly, more quietly, and with less urgency in the Blue Zones. The daily rhythm has been crafted such that life is less likely to be laced with worry, hurry and the constant need to be elsewhere. It is not a coincidence that people there live longer lives.

My recommendation to patients for downtime, stress reduction and relaxation is to have a period or periods each day when the phone and other internet communications are switched off. Do this regularly and acclimatise to switching off, gradually extending the time period. Let people know that you will be doing this. In that way, you will feel under less pressure when temporarily parted from your technology. If possible, leave your phone outside the bedroom door at night and try not to engage with it for up to one hour before bedtime. Furthermore, if you can get together each day with friends, you will certainly reap the rewards.

I love the Michael Jordan retort when asked about worries regarding forthcoming games. He said, 'Why would I think about missing a shot that I haven't taken?' No wonder he is renowned for his 'cool' temperament and seemingly unworried attitude going into play. I use this thought when trying to distract myself from worries. After all, the more we focus on problems, the bigger they loom in our psyche. Taking the focus off a problem deflates and diminishes it.

Sharing problems also reduces stress and worry. An experiment at the University of Southern California set out to test the well-known slogan 'a problem shared is a problem halved'. Participants were paired up and asked to make a speech while being taped by researchers. Prior to each speech, half of the participants were encouraged to discuss how they felt about public speaking. The remaining participants were told not to discuss their feelings. Stress levels were measured before, during and after experiments. Stress was significantly reduced in the participants who were able to vocalise how they felt and share their fears, worries and expectations.

Cortisol levels were also measured before, during and after each participant's speech. Cortisol is a hormone that is a strong biomarker of stress levels. Levels were significantly lower in the 'sharing' pairs. When cortisol remains chronically elevated, it accelerates inflammation and cell ageing. So, the best way to beat stress is to share feelings – and sharing them with someone in the same situation yields the best results. This is because sharing a threatening situation with a person in a similar emotional state buffers individuals from experiencing the heightened levels of stress that typically accompany threat.

———◆———

Gardening is consistently reported as one of the most popular hobbies. It is an excellent tool for helping with stress management as it's creative and requires considerable concentration, thus enhancing relaxation and well-being. Nothing slows us down like petunias and tomatoes. Spending time in nature reduces stress and results in a feeling of being in control. Engaging with a garden distracts from worries and stops us obsessing about problems. For generations, gardeners have known that planting, watering, weeding and all the beauty that emerges as a result is good for us. And science is catching on too. Numerous studies show that gardening improves physical and emotional well-being. Whether it's a few plants on the windowsill, containers on a deck, beds and borders in the yard or a vegetable plot, gardens big and small can be incredibly beneficial for the gardener and others who share in the results.

A recent review confirmed how gardening combines physical activity with social interaction and exposure to nature and sunlight. Sunlight lowers blood pressure as well as increasing vitamin D levels in the summer and the fruit and vegetables that are produced have a positive impact on diet. Working in the garden restores dexterity and strength, and the aerobic exercise that is involved can easily use the same number of calories as might be expended in a gym. Digging, raking and mowing are particularly calorie intense. The social interaction provided by communal and therapeutic garden projects for those with poor mental health can have remarkable health advantages. Furthermore, it has also been reported that the social benefits of such projects may delay the symptoms of dementia. Patients who are recovering from a heart attack or stroke find that exercise in a garden is more effective, enjoyable and sustainable for rehabilitation than therapy in formal exercise settings.

Another recent paper analysed 22 major studies on gardening and health. Health outcomes for people who don't participate in gardening were compared with those who do. The studies reported a significant positive effect for gardening in a wide range of health outcomes, such as reductions in depression, anxiety and body mass index, as well as increases in life satisfaction, quality of life and sense of community.

In a field experiment, the stress-relieving effects of gardening were tested by giving allotment gardeners a psychological task that put them under pressure. They were then randomly assigned to outdoor gardening or indoor reading, during which activity, stress hormones and mood were repeatedly measured. Gardening and reading each led to decreases in stress hormones and improved mood, but the degree of change was significantly more with gardening and improvements were sustained for longer, thereby providing experimental evidence that gardening promotes relief from acute stress and the benefits persist after the task ends.

In an experiment in clinically depressed adults, changes in the severity of depression and the ability to concentrate were measured during a 12-week therapeutic horticulture programme. Depression scores significantly improved during the gardening therapy and improvement was sustained after the study had ceased in three quarters of the cases. The improvement in symptoms of depression during the gardening therapy was driven by how much it captured the participant's attention. In other words, depressed participants had to like what they were doing to reap the benefit.

The world of gardening offers something for everyone, so whatever your preference – from virtual tours of botanical gardens to nurturing house plants, from allotment gardening to planting a few small raised beds, the evidence strongly supports engagement with gardening as a means of destressing and improving mood.

There may also be a biological reason why getting dirty outdoors is good for us. One type of bacteria that is common in soil has been found to stimulate the production of a mood-lifting hormone. This may go some way to explain why working with soil makes us feel so happy. The bacteria, Mycobacterium vaccae, triggers release of serotonin, the chemical principally responsible for elevation in mood and decrease in anxiety. An illustration of how important serotonin is for mood is that many of the medications used to treat depression act through serotonin brain pathways.

Being surrounded by greenery makes life more manageable, even in urban environments. Given that the natural environment can enhance mental and physical health, governments are starting to regenerate urban environments to be greener and more in tune with nature. Regeneration of nature in the form of wildlife gardens, for example, doesn't just benefit us – they encourage bugs, bees and birds, too. The environment benefits and the visible nature around us makes us feel less stressed.

The Japanese Ministry of Agriculture, Forestry and Fisheries coined the term *shinrin-yoku*, meaning 'taking in the forest atmosphere or forest bathing' and have embarked on a new programme of reforestation. Researchers in Japan conducted field experiments where subjects viewed either a forest or city area. Objective stress markers, including hormones, blood pressure, heart rate and other biomarkers of nervous system activity, were measured before and after walking and viewing. The forest environments lowered all stress markers significantly compared with walks in the city. Forests promoted lower levels of cortisol, whereas activity in the section of the nervous system which calms the heart rate and other body systems (parasympathetic) was enhanced. Conversely, activity in the section of the nervous system responsible for fight-or-flight responses (sympathetic) and stress reactions was reduced. All remarkably striking good news for forest walking and viewing.

Worldwide, similar research on forests and human health has led to new programmes for urban forest development. The International Union of Forest Research Organizations (IUFRO) established in 1892, with headquarters in Austria, is a not-for-profit international network of forest scientists who meet every five years to promote global cooperation in forest-related research. The IUFRO, representing over 15,000 scientists, fosters cross-disciplinary dialogue between professionals in the fields of forestry and health, as well as promoting international efforts in 'forests for bathing', including in urban environments. Growing research in the potential of the natural environment to enhance health and well-being accentuates the underuse of this resource as a health promotion tool.

———◆———

How many times in a week do you eat alone? I am well aware that many of us have no choice but to eat alone. Let's first review some of the implications and then solutions. Eating alone has many psychological and health disadvantages whereas sharing a meal with friends or family is a simple way to destress. One study measured the mealtime routines, challenges and preferences of people over 75 who lived by themselves. The biggest mealtime challenge was lack of the shared family experience, including lack of companionship. The preferences of these older persons was overwhelmingly to eat at least one meal a day with others. More than three-quarters of respondents said they wished their families shared more meals with them. One in five people over 75 are lonely when eating alone, three quarters eat alone most of the time and many people skip meals because they find eating alone too lonely. A majority eat more nutritiously and say food tastes better when eating with others. People spend more time eating when with others than when alone – on average 43 minutes compared with 22 minutes. A majority of older persons recalled that family mealtimes were important occasions for conversation and sharing when their children were younger. Seventy-eight per cent said they wished their families shared more meals together. Solo eating is not the exclusive domain of older persons. Nearly half of adults' meals are eaten in front of the computer, in the car, on the go – or, in other words, alone.

Let's take a step back and reflect on the benefits of shared meals and what we can do to change this epidemic of solo eating. Sitting down together at mealtimes is good for mental health at all ages. Whether it be through sharing experiences and bonding with family and friends, winding down with company or just having someone to talk to, mealtimes provide a great opportunity for us to set aside a specific time of the day or week to socialise, to relax and to improve mental well-being. Sharing meals develops social skills in children and adolescents who learn from behaviour modelled by grandparents, parents and older siblings. Mealtimes provide an opportunity where children and adolescents learn to listen and interact in conversation. Qualities such as empathy and understanding are developed, as views and perspectives other than one's own are discussed. Mealtimes are the perfect opportunity to share the invaluable wisdom that older adults have accumulated over a lifetime. 'Shared eating', with generations of family and friends coming together to share meals, is standard practice in all the Blue Zones and is cited as one of the reasons for the centenarians' healthy longevity.

In light of the evidence showing that eating alone or on the run increases the risk of obesity and poor nutrition, as well as negative inter-generational outcomes, a renaissance of the family meal or regular meals with friends is surely in order. The UK's Mental Health Foundation has made the following recommendations for shared meals:

Make a date – Set achievable goals. Set aside at least one day every week for sharing a meal with family or friends. This should be an event that is an honoured and routine part of every week, whether it's a leisurely breakfast, dinner on a Friday or lunch on Sunday. Make sure everyone is involved, both in deciding the day and in making sure it's kept free. I recommend that when a face-to-face meal is not possible, one tries to use technology to ensure that all or most members are present at one meal per day.

Hassle-free meals – When planning the meal try to choose something that is tasty but relatively simple and easy to prepare. This will ensure that the tradition continues and doesn't become a chore.

Share responsibility – Get everyone involved. Decide who will choose what will be served, do the grocery shopping, set the table, cook and do the dishes. Rotate these tasks.

Plan meals in advance – Planning meals in advance will save time in the long run and provide an opportunity to put a little more thought into introducing a variety of interesting dishes into mealtimes. Ask others for input into meal planning.

Involve children and grandchildren – Over the course of time, get them involved in all aspects of mealtime preparation, from menu planning to cooking to doing the dishes.

Telly-free – Try to use the opportunity mealtimes provide to talk and share. A television on during a meal is distracting, even if it's only in the background.

If none of the above is possible for you then try to adopt ways to improve mealtimes alone. Assemble a healthy appetising meal at least once a day, putting time into preparation whilst listening to your favourite podcast or watching a good TV show. Even try out new recipes that challenge you. Eat out more frequently. Bring a book when eating alone and enjoy the experience. If you have friends eating alone, share a call during joint mealtimes and even jointly experiment with recipes. Many of us have no choice but to eat alone so don't hesitate to reach out to others if you can because they will most likely be feeling the need as much as you are.

My colleague Shane O'Mara, a neuroscientist at Trinity College, Dublin, has written a bestseller about the many benefits of walking on mood and brain function. In his book, *In Praise of Walking*, he includes further evidence for the superior impact of walking outdoors in the natural environ. If we become used to walking and then stop, we miss the stimulus and become cranky and dissatisfied. Personalities change for the worst if we're deprived of walking.

With bodies in motion, we think more creatively, our mood improves and stress levels fall. Researchers from Stanford showed how walking boosts creative inspiration. They examined creativity levels while people walked versus while they sat. A person's creative output increased by 60 per cent when walking. Even walking indoors on a treadmill in a room facing a blank wall produced twice as many creative responses compared to a person sitting down and these responses were further enhanced by walking outdoors. The study also found that creative juices continued to flow when a person sat back down shortly after a walk. Walking and creativity are both stress busters and positive mood enhancers.

When the fight-or-flight response is appropriately invoked, it helps us rise to a sudden challenge. But trouble starts when the response is constantly provoked by stress and day-to-day events, such as money woes, traffic jams, health worries, job worries or relationship problems. So, having discussed the issues with and causes of stress, let's now reflect on evidence-based techniques to destress.

Relaxation and destressing responses through controlled breathing were first introduced in the 1970s by a Harvard cardiologist to mitigate against chronic stress. Slow, deep, regular breathing induces relaxation through enhanced parasympathetic activity. Simply breathing in slowly and deeply, pushing the stomach out so that the diaphragm is put to maximal use, then holding the breath briefly and exhaling slowly invokes relaxation. This should be repeated 5 to 10 times, concentrating on breathing deeply and slowly. Deep breathing is so easy and can be done at any time, in any place.

Rigorous scientific studies have confirmed the value of the ancient practice of meditation for not only dealing with stress but achieving long-term holistic health. Using brain scans, meditation has been proven to preserve grey and white matter, the main structural tissues of the brain. It also has the potential to suppress processes implicated in brain ageing and confer 'neuroprotection', which means protection of brain cells from decay and death. Meditation increases brain blood flow and oxygenation and decreases the action of the sympathetic 'fight-or-flight' nervous system, with a corresponding increase in parasympathetic 'relaxation' nervous system activity. Neuro-trophins, the family of proteins that increase survival and lifespan of brain cells, rise as a consequence. The mitochondria, present in every cell in the brain and body, produce 90 per cent of the cell's energy and this energy production increases during meditation. So, given these remarkable holistic benefits, having a go at mediation is surely a 'no brainer'!

Thích Nhát Hanh is a Vietnamese Thien Buddhist monk and strong advocate for mindfulness. He is, at the time of writing, aged 93. He has many wonderful sayings. With respect to mindfulness, he explains that, 'Life is available only in the present moment which underscores the principal behind mindfulness.'

Dispositional mindfulness (sometimes known as trait mindfulness) is a type of consciousness that has only lately been given serious research considerations. It is defined as a keen awareness of and attention to thoughts and feelings in the present moment, and research shows that the ability to engage in this has many physical, psychological and cognitive benefits, including decreasing stress and worry. Dispositional mindfulness is a quality in life, a fixed trait, rather than a state to enter into during practice.

Mindfulness requires training. We are constantly allowing our minds to wander, particularly into the future and to places of worry about possible forthcoming events. This causes us to fret about things that have not happened and may never happen, rather than focusing on the present. Distraction is unhealthy and a waste of time. Mindfulness is like a workout for the brain. We repeatedly pull our thoughts back to the present. This can be practised during a period of time each day or, better still, become a part of our everyday lives where we learn to 'stay in the present' at all times. The latter is dispositional mindfulness. Recently, there has been a surge of interest in how mindfulness and meditation improve biological ageing and, in particular, enhance the immune system. More trials are necessary to confirm these promising observations.

Another technique that I recommend to patients is progressive relaxation of skeletal muscles – the muscles we use to move our bodies and that are under our control, as opposed to the heart muscle, for example. Stressed muscles are tight and tense and by relaxing them, we dissipate stress. Muscle relaxation takes more time than deep breathing. It is best performed in a quiet, secluded place, comfortably stretched out on a firm mattress or mat. It focuses sequentially on the major muscle groups. Tighten each muscle and maintain the contraction for 20 seconds before slowly releasing it. As the muscle relaxes, concentrate on the release of tension and the sensation of relaxation. Start with facial muscles and then work down the body until you come to relaxing your toes. The entire routine will take 12 to 15 minutes. Initially it should be practised twice daily; expect to master the technique and experience some relief of stress in about two weeks.

Yoga is gaining increased popularity as a therapeutic practice with over 6 per cent of Americans being recommended yoga by a physician or therapist. Half of American yoga practitioners themselves reported starting practice explicitly to improve their health. In the United Kingdom, yoga is promoted by the National Health Service as a safe and effective approach, in health and illness, for people of all ages.

Yoga originated in India over 2,000 years ago. The term yoga is derived from the Sanskrit word *yuj*, meaning 'to join' and symbolises the union of body with consciousness. It combines physical postures, breathing techniques, relaxation and meditation.

Studies on yoga have increased 50 fold since 2014 and some of the most persuasive research relates to stress, insomnia and anxiety, as well as more established benefits to physical health conditions, including diabetes, hypertension and coronary heart disease. It is particularly good for improving balance and flexibility. It works through a mix of increase in positive attitudes towards stress, self-awareness, coping mechanisms, control, spirituality, compassion and mindfulness. At a cellular level, it reduces inflammation and thereby slows biological ageing. Yoga increases cannabinoid and opiate levels and affects nervous activity between the brain and the stress control glands (adrenal glands) on the kidneys, which release chemicals that relax blood vessels – all good stuff!

I discussed telomeres, the protective coverings at the end of chromosomes which stop chromosomal damage, in the first chapter. With ageing, telomeres shorten and because chromosomes are consequently damaged, cells decay and die. Telomerase is an important enzyme that prevents shortening of telomeres. In a number of studies, yoga affected telomerase and telomere length. One elegant paper from the All India Institute of Medical Sciences showed enhancement in telomerase and an increase in telomere length with yoga practice. Other important indicators of cell ageing that we previously discussed such as cortisol, endorphins, cytokines – and another that we will discuss later, BDNF – also change to a more youthful profile with yoga.

Taken together, evidence is growing that interventions such as yoga, meditation, breathing exercises and mindfulness improve the physiological biomarkers of cell ageing and therefore slow the ageing process. Coupling these with regular periods of separation from our devices and increasing the time we spend with nature will reduce stress and further decelerate biological ageing.

Chapter 7

In Search of the Elixir of Youth

THE THIRST FOR ETERNAL YOUTH HAS PLAGUED MANKIND for as long as we can trace. From AD 618 to 907, the Tang dynasty represented the most prosperous period in China's history. Chinese culture flourished during an era considered the greatest age for civil society and particularly for Chinese poetry and art. The system of government involved competitive civil service-type posts, ensuring academically able advisors to the dynasty. Emperors were obsessed with immortality and the search for the elixir of youth and yet, despite high levels of sophistication and culture, remarkably 6 of 22 of the Tang emperors died from accidental self-poisoning in the pursuit of eternal youth. According to alchemists of the dynasty, 'blood-red cinnabar', 'fickle volatile mercury', 'gleaming gold' and 'fiery sulphur' were the principal ingredients for immortality. They are also deadly poisons responsible for the emperors' deaths as well as those of noblemen who perished in the bid to achieve perpetuity. Emperors and noblemen were not alone in their obsession, which fascinated scholars and statesmen alike. The famous Chinese poet Po Chu-I spent hours bending over an alembic, stirring concoctions of mercury and cinnabar. However, for some unknown reason, he did not imbibe himself and as a result outlived his unwitting friends and family, writing:

> 'At leisure, I think of old friends,
> And they seem to appear before my eyes…
> All fell ill or died suddenly;
> None of them lived through middle age.
> Only I have not taken the elixir;
> Yet contrarily live on, an old man.'

I wonder when the penny dropped for Po Chu-I? For others, it took almost 300 years before the connection between the concoctions and deaths were realised and the practice of drinking the alchemists' concoctions was discontinued. I was reminded of this Chinese story when a US president made reference to the possible benefits of drinking disinfectant as a means of killing the SAR2COV virus. Thankfully, these days, we have learned to be much more discerning and exacting when it comes to prescribing and imbibing.

Fast forward from the Tang dynasty to the 21st century and to Larry Page, the co-founder and former CEO of Google, who has established a company to pursue the search for a 'cure' for ageing. In 2013, Google launched Calico, which, the company's website states, is 'tackling ageing, one of life's greatest mysteries'. As part of this ambitious and costly venture, Calico has invested in research in many different areas, including into an interesting and strange small mammal, the naked mole rat, which, despite its small size, has an unexpectedly long life.

The naked mole rat is the size of your middle finger. Not a handsome mammal, it is a small, hairless (hence 'naked') and blind rat that lives underground in east Africa. It has two long protruding hooked teeth which resemble fangs and move independently of each other. The rat tunnels its way underground by using these two front teeth and can withstand low oxygen concentrations that would kill all other mammals. For example, human brain cells start to die within 60 seconds of oxygen starvation and permanent brain damage typically sets in after three minutes. In contrast, the naked mole rate can survive a full 18 minutes in an environment that has no oxygen without suffering any harm to brain cells or other cells. So, from our scientific perspective, this mammal could provide new treatments for brain damage caused by strokes if we could determine how it withstands such long periods without oxygen. As well as surviving extremes, it lives for 30 years, it never gets diseases of ageing such as cancers or heart disease and does not die from old age as we know it. Death occurs as a result of attacks by other animals or sometimes infection.

The queen mole rat, with the assistance of her cohort of males, amazingly maintains a consistent rate of fecundity and she has no menopause – something else which is of interest to scientists. Readers who have struggled through a menopause will appreciate the import of this! In addition, blood vessels remain in good condition throughout life, with negligible loss of elasticity and no 'hardened arteries' as is so often the case in other ageing females after the menopause and of course in older men. So, is it possible that, in contrast to the ancient Chinese formula of cinnabar, mercury and sulphur, the recipe for today's formula for the 'elixir of youth' may reside within this obscure small, unattractive mammal, under investigation by Calico?

The naked mole rat – sugar cubes to indicate scale. (Jane Reznick/Gary Lewin, MDC)

It is disheartening to learn that more than 99.9 per cent of species that have ever lived are now extinct. But, even with this notable statistic, an estimated 10–30 million distinguishable species currently inhabit the earth. The 'life sciences' is the term used for the study of these species. Most of what we know about ageing comes from the combined study of the life sciences, including the biology, medicine, anthropology and sociology of different species. It is humbling to know that the genesis of the contribution of the life sciences to our understanding of why we age dates back over four centuries, when it was realised that how we age exhibits common characteristics across all species.

Georges-Louis Leclerc was the polymath to whom we owe this observation. Leclerc himself has an interesting life story. In the early 1700s, after a Jesuit education in Dijon, France, he studied law. He then went on to study mathematics and eventually medicine. After completing his medical studies, he was lucky enough to inherit a veritable fortune. This left him free to pursue his ambition for a life of science without any pressure to generate an income. Although not a biologist by education, he described the evolutionary theory of biology, which states that 'the ageing of species is common to all species'. The implications of his observation were far reaching and, as a consequence, biologists today can study genes relevant to ageing in the housefly and apply research findings to human studies because the genes are the same in both species. There is a lovely old Irish expression, '*Cad é a dhéanfadh mac an chait ach luch a mharú?*' which translates as, 'What does the son of a cat do but kill a mouse?' Leclerc was one of the first to imply that we get inheritance from our parents, in a description based on similarities between elephants and mammoths. Except for Aristotle and Darwin, no other student of the natural world has had as far-reaching an influence and yet we are all familiar with Aristotle and Darwin but less so Leclerc.

Latter-day work on the biology of ageing confirms Leclerc's impressions and provides important clues about how to develop effective interventions that delay ageing. It is now clear that some of the hormonal and cellular pathways that influence the rate of ageing in lower organisms, such as flies or worms, also contribute to many of the manifestations of ageing that we see in humans, such as cancers, cataracts, heart disease, arthritis and dementia. Several studies have demonstrated that by manipulating certain genes, altering reproduction and reducing caloric intake, the duration of life of both lower organisms and mammals can be extended. Lower species are easier to study in large numbers, particularly drosophila, the common housefly. I have visited many laboratories that house large glass containers full of noisy buzzing flies, which are the mainstay of the laboratories research activities into ageing. Much of our knowledge of why human cells age derives from observations in such lower species. Perhaps next time we go to swat a fly, we will pause and reflect on its contribution to science!

We humans are very advanced organisms – you have been working on 'you' for millennia. You exist only because of billions of deaths of organisms that were less well-adjusted and less complex. You are the survivor – you exemplify 'survival of the fittest'. You started off your journey 4 million years ago as a single cell. Today's cells have altered little in core content from that first cell. Cells are very, very tiny. For example, it takes 10,000 human cells to cover a pin head and our bodies are made up of trillions of cells.

A cell's main job is to produce energy, the energy that keeps the cell and therefore us alive. Very simply, food is converted into energy by the cell; waste occurs as a by-product of this energy and is rapidly disposed of by the cell. The instructions for energy creation and for waste disposal come from the cell's nucleus. As we learned earlier, the nucleus is the cell's library – a digital library that holds all of the cell's information and sends out instructions through the cell at regular intervals when required. The cell wall allows the toxins and waste, which are the by-products of energy production from metabolism of food, to leave the cell – and eventually the body via the intestines and bladder as faeces and urine – and keeps all the good chemicals and food within for energy production. So, anything that changes the strength of the wall can cause serious damage. The part of the cell that produced energy and is responsible for the energy transactions necessary for cell survival is the mitochondria.

Our cells are constantly busy, they never rest, generating energy and dividing and producing new cells repeatedly. During these divisions, the genes also divide and thereby pass on the instructions for various characteristics to the next generation. Occasionally, there are imperfections in divisions, called mutations. A mutation alters the instructions for one or more characteristics. Some mutations are minor and we live with them unnoticed but many mutations lead to death or impairment of the organisms. In this way, superior organisms, such as you and me, slowly evolved towards greater complexity – we are the survivors.

Individual cells have a finite lifespan, so when they die they are replaced by new cells, which is why all division or replication is so important to us. Cells are dying and being replaced all of the time. Anything which interferes with this delicately balanced cycle between cell death and cell reproduction will hinder replacement of new fully functioning cells and thereby contribute to ageing of the organism.

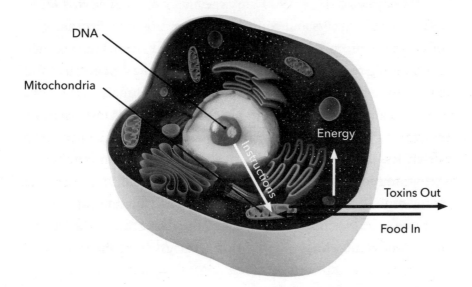

The makeup of a human cell

Each type of cell has its own lifespan – this is important in forensic science and in murder investigations. For instance, red blood cells live for four months, white blood cells for a year, skin cells for three weeks, colon cells for four days and sperm cells three days. By knowing cell lifespans, time of death can be surmised depending on which cells are still living.

In natural animal populations, predation, starvation and environmental stresses rapidly eliminate frail or old animals. Humans are the striking exception to this rule, achieving a life expectancy of over 80 years despite frailty and old age. In the last 200 years, the average human life expectancy has doubled in most developed countries. In a century, the world has changed markedly from there being almost no countries where the life expectancy of the average citizen was of 50 years to many countries having a life expectancy of 80 years or more, and the pace that these changes are occurring is extraordinary. The following title made the cover of *Time* magazine in 2015: 'This baby could live to be 142 years old'. In 1900 life expectancy of females was 47. In 2010, it was 79 years and continues to rise. You may well ask why this is happening. We don't have all of the answers but extended lifespan is generally due to the ability of human populations to manipulate their environment, domesticate animals and plants, use tools and fire to provide stable nutrition and almost eradicate parasites. All of this is coupled with advances in medicine, clean water, less stress, more prosperity and, of course, our dominance over mutations as we evolved. The biological consequences and population impact of ageing is thus a uniquely human experience. For example, women will live half of their adult lives after losing reproductive potential – a situation unheard of in the rest of the mammalian world.

We can learn more about what else is contributing to this human longevity from some animal species who live extraordinarily long lives. For most animals, there are two basic ways to die: from ageing and thus disease, and from injury. But a select few species are seemingly immune to ageing or diseases. In these animals, the gradual accumulation of cell damage that eventually kills most cells is slowed down dramatically – almost to a virtual standstill – thus prolonging life and youth. This is

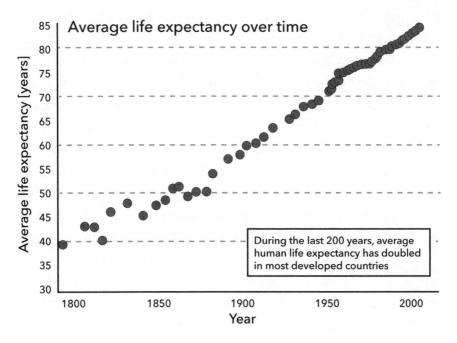

Average life expectancy in European countries with accurate death records since 1800

referred to as 'negligible senescence'. It is fascinating how long some of these species live for. Tortoises are the most famous negligibly senescent animals. When an Aldabra giant tortoise named Adwaita died in 2006, carbon dating of his shell confirmed that he was born around 1750 – he was 255 years old. He died of liver failure complicated by a wound brought on by a crack in his shell. If his handlers at the Alipore Zoological Gardens in Kolkata, India, had had the resources and inclination to arrange for a liver transplant and surgery to rebuild his shell, Adwaita would still be crawling around today. Even so, making it to 255 years old wasn't bad!

The Antarctic sponge has extraordinary longevity, living up to 1,550 years. It hardly moves at all, leading to one of my colleagues cruelly referring to a particularly lethargic member of his team as 'the sponge'. The bowhead whale, the largest of long-lived mammals, lives over 200 years (the oldest known whale was 211 years). The jellyfish Turritopsis nutricula is the most fascinating of all, transitioning from polyp to adult to polyp, thereby displaying 'eternal youth' – the Benjamin Button of the animal world. Spare a thought for the queen termite with a lifespan of 50 years – the poor thing produces 30,000 eggs per day. Contrast this with the female American mayfly, much beloved of fly fishermen, which lives for only five minutes. You can imagine how these extremes in lifespan are of interest to us scientists in gerontology – what is it about the cell functions of different species that drives longer, or shorter lifespan? Were we to understand this and develop the ability to mimic the relevant changes in human cells we would likely be able to slow down ageing and occurrences of disease and thereby extend healthy human life – the much sought-after 'elixir'.

◆

OK, ladies, take a bow, because women live longer than men in almost all modern societies. Longer life in females is not just the case for humans but also for many other mammals, such as chimpanzees, gorillas, orangutans and gibbons, all of whom consistently outlive the males of the group. On average, women live between six and eight years longer but this gap is narrowing in western civilisations, mostly because deaths from cardiovascular diseases are declining in men.

There are a host of other plausible explanations for the gender gap, including biological, hormonal, genetic, environmental and social factors, all of which likely contribute in varied proportions. One popular biological explanation is the gender difference in metabolic rate. Metabolic rate, the amount of energy that we produce from metabolism of food, is about 6 per cent higher in adolescent males than females of the same age and increases to 10 per cent after puberty. In many experiments and in most species, metabolic rate is negatively associated with longevity – that is, high metabolic rate is coupled with a reduced lifespan.

Females also tend to convert more food into white adipose tissue compared to males, who convert their food into muscle (a good thing) and circulating lipids including LDL-cholesterol (bad cholesterol). Cholesterol is one of the leading risk factors for cardiovascular disease. The female hormone oestrogen has protective cardiovascular benefits, reducing the levels of LDL-cholesterol and increasing HDL-cholesterol (good cholesterol), thus protecting pre-menopausal women from heart disease. Oestrogens further protect the inner lining of blood vessels from injury, dilate blood vessels and thus lower blood pressure and protect against atherosclerosis by reducing both clots and hardened arteries. All of these factors constitute women's superior cardiovascular profile and longer lives. Finally, in some countries men are also more frequently exposed to occupational hazards – they drive more miles, drink more alcohol, smoke more and are more often exposed to trauma, including homicide. Although we ladies in the Western world live longer, as societies evolve, this gender gap is narrowing and men who adopt better health behaviours can further reduce the difference – where there is the will to do so.

A well-known scientist in this field, Caleb Finch, when asked how long could we live if we didn't age or did so at the turtle's rate, reflected that, 'In theory, if mortality rates did not increase as usual during ageing, humans would live hundreds of years. I have calculated for humans that at mortality rates of 0.05 per cent per year, as found at age 15 in developed countries, the median lifespan would be about 1,200 years.' Of course, what happens is that mortality rates accelerate as we get older – they do not stay constant at the rate they were when we were aged 15 in contrast to negligible senescent animals. As Finch explains, negligibly senescent animals enjoy a 1–2 per cent death rate once they get past their seventieth year in contrast to human mortality rates, which accelerate past 70 years. At 65–70 we have a 1:100 chance of dying in the next five years, this rises to a 1:10 chance after age 85. Compare this with 1:10,000 for six-year-olds. At present, the reasons why some animals have negligible senescence are not clear. It may be an evolutionary development which gives them a reproductive advantage, or simply an accident. There is much research activity in this space which could hold the answer to the 'elixir of youth' or the 'elixir of life' for humans.

A more realistically achievable goal than negligible senescence in our lifetime is a modest deceleration in the rate of ageing sufficient to delay all ageing-related diseases by about seven years. This target is chosen because the risk of death and most other negative attributes of ageing tends to rise exponentially throughout the adult lifespan with a doubling time of approximately seven years. Such a seven-year delay would yield health and longevity benefits greater than what could be achieved with the elimination of cancer or heart disease. If we succeeded in slowing ageing by seven years, which scientists consider a realistic target, people who reach the age of 50 would have the health profile and disease risk of a 43-year-old; those aged 60 would resemble 53-year-olds, and so on. Equally important, once achieved, this seven-year delay would yield equal health and longevity benefits for all subsequent generations, much the same way that children born in most nations today benefit from the discovery and development of immunisations. I believe that this is an achievable goal and many of the elements covered in this book – friendship, stress relief, laughter, purpose, sleep, foods, physical activity and positive attitude do exactly that – delay age-related diseases and disorders and death by up to seven years or more. The earlier we address risk factors which influence the ageing process, the more reserve we can build up in body and brain capacity and the more likely it is that we will achieve a seven-year delay in ageing.

Knowing why some animals have particularly long lifespans should inform how we might be able to manipulate cell functions or structures to enhance longevity and reduce age-related diseases.

Chapter 8

Cold Water and Hormesis

THE NEXT TIME YOU GO TO A FANCY SPA AND ROTATE BETWEEN exercise, sauna, steam room and cold-water pools, take a moment to pause and reflect on how ancient these rituals are. People have derived pleasure from bathing and water for around 4,000 years. Public baths existed in early Egyptian palaces as early as 2,000BC and bathing also occupied an important place in the life of the ancient Greeks. However, we know much more about the high degree of sophistication of Roman baths, thermae – a complex of rooms designed for public bathing, relaxation and social activity – just like our modern day spa.

The Roman technique of bathing followed a somewhat standardised pattern. The bather first entered the apodyterium, where he undressed. He was then anointed with oil in the unctuarium before entering a room or court, where he indulged in rigorous exercise. After this activity, he proceeded to the hot room and then the steam room where his skin was ridded of its accumulation of oil and perspiration. The bather then moved to the tepidarium (warm room) and afterwards to the frigidarium (cold room), where there was frequently a cold water swimming pool. The bathing process was completed after the body was once more anointed with oil. What a pleasant way to pass a few hours.

The use of water for therapeutic purposes is an ancient practice. These days, water therapy is used to treat musculoskeletal disorders such as arthritis or spinal cord injuries and prescribed for patients suffering from burns, stroke or paralysis. As with the Roman bath, integral to the bathing experience is cold water exposure. There is copious evidence for the health benefits of cold water on a number of systems and pathways involved in the process of ageing.

Cold water immersion provides a stimulus to our physiological systems which is related to the phenomenon of hormesis, whereby small amounts of a harmful or painful agent are actually good for us. The work on this counterintuitive phenomenon – in which moderate doses of stressors such as cold exposure, radiation, noxious compounds and starvation prove beneficial rather than harmful – intrigues gerontologists. Laboratory organisms often survive longer after exposure to such stressors and we are of course eager to learn whether and how we can harness this response to enhance cell life. What we do know to date is that exposure of cells to mild stress stimulates the synthesis of proteins, which improve the function and survival of cells without interference in their ability to reproduce and divide. We believe that the reason is that the triggering of one recovery mechanism in the cell improves the functioning of other repair and recovery systems. Whatever the reason, hormesis is fascinating. We can use this to explain why cold water exposure is so good for us and good for ageing.

A cold shower or cold water immersion is a physiological stress which has a hormetic effect as it forces the body to recover to normal core temperature after the cooling stimulus, resulting in indirect benefit to many systems and organs. Other moderate stressing agents could therefore theoretically be beneficial too – for example, hypoxic stress (holding one's breath), oxidative stress (hyperventilation) and heat shock (sauna), but these have not been tested to the same degree as cold water, in the context of ageing.

Cold water immersion or cold water showers are so effective in delivering a large-scale stimulus to the body because the number of cold receptors in the skin is up to ten times more than warm receptors. Furthermore, the ability of water to conduct temperature is 30 times stronger than air. On exposure of the skin to cold water, blood vessels contract and raise blood pressure, which, coupled with the temperature shock, causes electrical impulses from peripheral nerve endings to travel to sensory centres in the brain. The result is an increase in important chemicals and nerve signals. One of these chemicals is noradrenaline, a critical neurotransmitter that is part of our 'flight-or-fight' response, which rises four-fold on cold water exposure. Noradrenaline boosts performance of cells in both the brain and body and regulates a host of functions such as heart rate, blood pressure, blood flow to muscles, power of contraction of skeletal muscles and release of energy. Cold exposure also releases noradrenaline in the major brain areas that control emotions, concentration and memory, thereby affecting how alert we are, how good our memory is, our levels of interest in things, our mood and how the body responds to pain. Almost all of our organs use noradrenaline and the functions that it contributes to are central to the ageing process. Because responsiveness to noradrenaline declines with age, any stimulus which enhances its activity is important to 'ageing' physiology. One of my neuropsychology colleagues at Trinity College, Dublin has hypothesised that stimuli that excite the release of brain noradrenaline, such as cold water, may prevent dementia.

Noradrenaline is one of the chemicals involved in the sympathetic nervous system – the system that 'primes' the body for action. For example, the alertness we experience first thing in the morning is due to a surge of output from sympathetic nerves, which control blood flow throughout the body, mostly through increasing noradrenaline release. Cold water exposure also releases other chemicals, such as endorphins, which cause the familiar 'runner's high'. Cold water exposure causes a four-fold rise in endorphins, with consequent enhanced well-being and suppression of pain through stimulation of opioid receptors, further adding to the 'feel good' factor. Have you ever gone for a swim in the Atlantic and, though shivering going in and frantically splashing yourself to acclimatise, finding yourself elated, warm and glowing when you leave the water? Now you know why.

Cold water exposures improve immune responses. Ask any cold water swimmer or showerer and they will attest to having far fewer winter colds and chest infections and fewer illnesses overall. One study in which four groups of people were compared over many months supports these claims. The first group took hot-to-cold showers; the second did regular physical exercise; a third combined hot-to-cold showers and physical activity, whereas the fourth group did not change their behaviour. Compared to the fourth group, hot-to-cold showers resulted in a 29 per cent reduction in sick leave; physical activity resulted in a 35 per cent reduction in sick leave and the combination of physical exercise and routine cold showers resulted in a remarkable 54 per cent reduction in sick leave. The duration of the cold shower did not appear to make a difference. Participants reported an increase in energy levels, including many saying it was like the hit one gets from coffee. A further positive outcome was the improvement in quality of life, which was much higher in the cold water groups. Even though the vast majority reported a variable degree of discomfort during cold exposure, the fact that 91 per cent expressed the will to continue the routine after 90 days is perhaps the most indicative of any benefit.

One of the big effects of cold water swimming is its impact on how many calories we burn when at rest. When we swim outdoors, the body must work hard simply to stay warm and consequently burn more calories. The colder the water, the harder the body works to convert fat to energy. When this is coupled with the exercise of swimming, the calorie burn is further increased. The same chemical and sympathetic nervous system changes for cold water showers apply to cold water swimming. So, when we swim in cold water, the extreme change in temperature increases the sympathetic nervous system activity and reduces blood flow to the skin, both of which result in harder pumping of blood around the body by the heart to body organs, prioritising important organs such as muscle, brain and kidneys. As a result, circulation is improved and toxins are more readily flushed out of our system.

The 'clean skin and healthy glow' of regular cold water swimmers can be attributed to these physiological consequences. There is also good evidence to support associations between cold water swimming and decreases in tension, fatigue, improvement in mood and memory, and in general well-being.

Sound evolutionary theories support the reasons why we find cold water exposure so invigorating. Humans' lifestyle was mostly outdoors, with widely varying ambient temperatures, and frequent swimming or immersion in water of uncomfortable temperatures necessary to find food or escape predators. But, even though humans are homeotherms (our body maintains a constant core temperature of approximately 36.6°C), in modern life our utilisation of this sophisticated regulatory system is very low.

And this is not a good thing as the system requires stimulation. The rapid disappearance of thermal stress from the lifestyle of humans in the last few thousand years, compared to its presence for millions of years in primates and hundreds of thousands of years in homo sapiens, has had a negative effect on human physical and mental health because the thermoregulatory system does not get enough 'exercise' or stimulation. Consequently, cold water exposure is hard-wired; it triggers this evolutionary response and is invigorating.

———

The literature on the role of cold water in the treatment of depression is extensive and longstanding. The following case report from the *British Medical Journal* is a good illustration of how cold water swimming alleviated depression in a young female:

> 'A 24-year-old woman with symptoms of major depressive disorder and anxiety had been treated for the condition since the age of 17. Symptoms were resistant to well-known antidepressant drugs. Following the birth of her daughter, she wanted to be medication-free and symptom-free. She commenced a programme of weekly open (cold) water swimming. This led to an immediate improvement in mood following each swim and a sustained and gradual reduction in symptoms of depression, and consequently a reduction in, and then cessation of, medication. On follow-up a year later, she remained medication-free.'

Depression and low mood are common with advancing years. This is mostly because of changes in life circumstances, such as loss of a partner or job, or intrinsic factors, such as age-related neurotransmitter changes. It is well recognised that the noradrenaline system does not function as it should in people who have depression. Cold water immersion regularises the system and thus helps with the symptoms of depression. This is so for young and old. Patients enquire about the safety of cold water exposure, particularly in relation to heart attacks. It is the case that people with known heart disease should not embark on sudden temperature changes without medical clearance. The sympathetic surge could induce a heart attack if vessels to the heart are already narrow because of atherosclerosis or clots. Otherwise, brief whole-body exposure to cold water (15–23°C) is safe and without significant side effects, either short-term or long-term. The effect on core body temperature is so negligible that hypothermia is hardly ever a concern unless exposed for excessive long periods to cold water.

One further advantage of cold water exposure that is often overlooked but important to mention is the effect on skin. We talked about cold water causing a fresh, glowing look but it is also helpful for a well-known skin disorder that becomes more common with age – age-related pruritus, or itchy skin. With ageing, the skin struggles to retain oil and moisture and becomes dry. This results in itching and scaly red patches (the medical term for this is asteatotic eczema). Hot water showers and frequent hot water bathing aggravates or even causes asteatotic eczema. Cold water showers help to relieve the itching and do not cause dry skin to the same extent as hot showers.

—◆—

We can't leave the topic of cold water without reference to the sea. Maps of the world's population show that the bulk of humanity live near water. We live along coastlines, around the rims of bays, up the course of rivers and streams and on islands. We also vacation at the beach and find solace fishing on lakes. Nothing makes small children happier than the chance to splash through puddles. This human penchant for water makes evolutionary sense. When humans separated from apes and emerged from the forests of Africa, they stuck close to rivers and beaches, feasting on fish, clams and crabs. The marine diet was packed with omega-3 fatty acids – essential fatty acids that promote brain cell growth. Human brain growth began to increase exponentially thereafter.

Living near blue spaces – i.e., in proximity to the sea – is linked to better mood, less depression, less anxiety and better overall well-being. This applies to all age groups and some studies suggest that it is particularly so as we get older. Remarkably, proximity to the sea can add four to seven years to lifespan. Most of the science on extended lifespan and the sea comes from the Blue Zones, all of which are on high terrain and by the sea but also have a number of other characteristics that influence longer, healthier life – for example, good diet, active communities, less pollution and good quality of drinking water. So, teasing out the independent contribution of living by the sea from all of these other factors is challenging. The fact that stress and depression are less common may also be contributory.

Our research has shown that positive benefits on mood and well-being are more evident with the more visual exposure to the sea. In other words, being able 'to see the sea' matters. This applies to all ages and some studies show that it is particularly evident as we get older. The sea is constantly changing and it is well known that variety enhances well-being. The sea is never the same on two consecutive days or even for a couple of hours within the same day. Thus a sea view is never boring, always stimulating. Proximity to the sea also increases the likelihood of engaging in physical activity, such as swimming (even cold water swimming!) and walking. Seaside living may also increase social interaction and provides a space for health and well-being – all well proven to increase lifespan. Whatever the reasons – and they are likely to be multiple – the benefits from being near the sea are strong and equivalent to the impact that factors like wealth have on health and added life years.

Whether we adopt cold water showers, cold water swimming or just spend time watching the sea, it is well established that this will benefit our health and well-being.

Chapter 9

Eat to Your Heart's Content

THE FORBIDDEN MEAL, EATEN IN SECRET IN THE MIDDLE of the night, is one of my most cherished boarding school memories – the midnight feast. Usually, on a Saturday night, one of us was delegated to stay awake, check the coast was clear and rouse the other students at midnight, when we'd all sneak down to a large space under the staircase and proceed to share treats. These generally consisted of peanut butter and jam sandwiches and chocolate biscuits washed down with lemonade – nothing especially haute cuisine but mana from heaven for excited hungry schoolgirls. We were never discovered – the cat's out of the bag now of course – and I have happy recollections of the pleasures of nocturnal feasting.

However, given the intervening accumulated wisdom, I'm now going to put a bit of a dampener on midnight feasts or even any between-meal snacking, which is a recipe for poor health for most of us. The human body, through thousands of years of evolution, is hard-wired to consume as much food as possible for as long as it's available. In our early history, people hunted or gathered food and had brief periods of plenty, such as after a kill, and then potentially lengthy periods of famine. As humans were prey to large animals, they actively sought food during the day and sheltered and rested at night – so no midnight feasts.

Prior to the advent of our electricity-powered society, people started the day at dawn, worked all day, usually doing manual labour, and then went to sleep with the setting of the sun. Human activity was synchronised to day and night. This allowed for natural control of over-consumption. Today, we are working, playing, staying connected and eating day and night. This adversely affects body clocks, which evolved to operate on a sleep–wake cycle, conditioned to daytime activity, moderate eating and night-time rest. Yet we find snacking on sweets and sugars appealing because it's part of how we have evolved.

High calorie foods trigger dopamine release in 'pleasure centres' in the brain. These pleasure centres are linked by pathways that regulate biological clocks and physiological rhythms. Disruption of the pathways by eating high calorie foods, such as peanut butter and jam sandwiches between meals or during normal resting hours, results in the excess calories being stored as fat much more readily than if the same number of calories were consumed during normal feeding periods. The result is obesity and obesity-related diseases, such as diabetes and heart disease. As we get older, our sleeping patterns are disturbed and this may increase the likelihood of some of us heading to the kitchen at night to compensate with a quick snack. However, this will accelerate weight gain and it will not help with sleep. Bottom line: keep food intake to within an eight-hour window during the day if possible.

What has this to do with ageing? Food consumption, diet, genes and pathways related to metabolism and cell energy production are the most important controllers for how our cells age. We eat so that we can have energy. Food produces energy. The rate at which energy is used up by the body is referred to as the metabolic rate, which is a series of chemical processes in each cell that turn the calories we eat into fuel to keep us alive. There are three main ways our body burns energy each day. The basal metabolism refers to the energy used for our body's basic functioning while at rest. Next is the energy used to break down food and finally the energy used in physical activity.

An underappreciated fact about the body is that our resting metabolism accounts for a huge amount of the total calories we burn each day. Physical activity, on the other hand, accounts for a smaller part of our total energy expenditure – between 10 to 30 per cent (unless you're a professional athlete or have a highly physically demanding job). Digesting food accounts for about 10 per cent.

Two people with the same size and body composition can have different metabolic rates. We all have friends who fit into these categories – one can consume a huge meal and gain no weight while the other has to carefully count calories to avoid gaining weight. Whereas we don't fully understand the mechanism that controls a person's metabolism, we do know that the amount of lean muscle and fat tissue in the body, our age and our genetics change metabolism. We can change lean muscle mass and fat tissue but the other factors are of course fixed.

Metabolism slows down as we age. This age effect starts early – at age 18 – and continues throughout life, so that at age 60 we burn significantly fewer calories at rest than at age 20. Consequently, we are more likely to put on weight and to develop an important new syndrome referred to as the metabolic syndrome, which is a set of conditions including increased blood pressure, high blood sugar, a large waistline and abnormal cholesterol or triglyceride levels. With metabolic syndrome, we are at an increased risk of chronic health issues, including heart disease, stroke and diabetes.

Again, how this works, why these conditions cluster together into the syndrome in some people (estimated to be 30 per cent of people over 60 in our and other studies) and why it affects some people more than others remains unclear.

Why our energy needs go down as we age even if we keep everything else pretty much the same is also a mystery. Basal metabolic rate – the number of calories required to keep the body functioning at rest – is calculated by putting height, weight, age and gender into an algorithm online which uses equations summarised from statistical data. While there are certain foods, like coffee, chili and other spices, that may speed up the basal metabolic rate up just a little, the change is negligible and short-lived and would never have an impact on waistline. Building more muscles, however, can be marginally more helpful. The more muscle on your body and the less fat, the higher your metabolic rate. If your metabolic age – calculated by comparing your basal metabolic rate to the average of your chronological age group – is higher than your actual age, it's a sign that you need to improve your metabolic rate.

Basal metabolic rate is closely linked to the size of an animal and to heart rate and is widely accredited for determining how long animals and possibly humans live for. In general, the smaller the animal, the faster the basal metabolic rate and therefore the shorter their lifespan, although there are exceptions to this rule such as the naked mole rat. Smaller animals have a higher surface-to-volume ratio. Or, in other words, a relatively larger area for heat loss to the environment per unit time. It is imperative that animals (including us) keep a constant core body temperature for organs to function and survive. To keep a constant body temperature, a small animal must oxidise food and produce energy at a high rate to maintain body temperature. One of the smallest mammals is a shrew, a distant relative of the elephant. They weigh only about four grams. These animals have a metabolic rate that is too high to allow them to live much more than 12 months. To maintain this metabolic rate, they have a very fast heart rate of 600 beats per minute (compared with humans, at 60–80 beats per minute) and they eat nearly their own weight in (mostly) insects every 15 minutes to sustain themselves. They will starve to death in just a few hours if deprived of food and rarely stop to sleep. Because of the constant need for food, the shrew possesses a venom that paralyses its prey, keeping it alive for up to 15 days. The shrew then moves it to a cache. How shrewd, you might say.

There are exceptions to the rule that couple body size, basal metabolic rate, heart rate and life expectancy – these make for a tantalising area of ageing science. For example, rats and pigeons are pretty much the same size and have the same basal metabolic rate but pigeons live seven times longer than rats. The reason for this difference is that leakage of toxins and waste products during the creation of energy by the mitochondria is much less in pigeons, despite having the same metabolic rate as rats. As you can imagine, if researchers could understand why pigeon mitochondria are more leak-proof, then we might be able to use the information to modify human cell leakage and waste accumulation, which is key to ageing. If we crack this, could we apply the knowledge to human cell ageing and thus live seven times longer? That's a possibility that would shake our world big time.

Obesity is closely related to mitochondrial function. The obesity epidemic is accelerating and is now global; in fact, obesity is rising faster in middle-income countries than the West. In our study, we found that 70 per cent of people over 50 in Ireland are overweight or obese. This is similar to data from other European countries, although Ireland and the UK are top of the obesity league compared with other European jurisdictions. Look around you – how many of your friends are at their appropriate body weight? The big issue with this is that with obesity comes accelerated ageing and early presentation of diseases – sometimes by as much as 20 years early – such as heart disease, high blood pressure, arthritis, liver diseases and skin problems.

The basal metabolic rate of overweight and obese people is higher than people of normal weight although the BMR per kg of body weight is lower. As with animals, heart rate is also higher to keep up with metabolic rate. This is one of the factors contributing to poor health outcomes from obesity. Obesity, a disease of excessive fat deposition, is essentially a consequence of chronic energy imbalance, with energy intake consistently exceeding expenditure. In other words, we eat more than we burn. This leads to the storage of surplus energy in white fat. We need to be much better educated about what fat is and how the body controls it if we are to get on top of the obesity epidemic, particularly as we age and body fat increases.

It's easy to lump 'fat' into a single category: the stuff beneath the skin that makes our stomach wobble and can raise the risk of diabetes and heart disease. But not all fat is created equal. For years, scientists have known that fat tissue comes in at least two different shades. White fat, which most of us are familiar with, stores energy in big oily droplets throughout the body and, in large quantities, causes obesity. Brown fat contains both smaller droplets and high amounts of iron-rich mitochondria that lend the tissue its chestnut colour. Green tea, cabbage, berries, spinach, hot peppers and coffee are examples of foods that increase brown fat production. Mitochondria, the cell's power plants where energy is produced, use these fatty droplets to generate heat. Brown fat is turned on when we get cold and is of interest because it can regulate conversion of fat into fuel or energy. Exercise can also stimulate hormones, such as irisin, which activate brown fat and thereby trigger energy release. So, to all intents and purposes, brown fat is good fat. Scientists are exploring new ways to harness this tawny tissue and irisin for therapeutic purposes to use in weight reduction because of the potential to turn fat into energy. White fat can turn into brown fat by exposure to lower temperatures of 19 degrees or less for a couple of hours a day. This may also be a further reason why cold water exposure, including cold showers, is beneficial.

While the solution to obesity is ostensibly as simple as reducing calorie intake (e.g., avoid energy-dense food) and increasing energy expenditure (increase physical activity), the failure of decades of public health initiatives to remove the 'obesogenic' environmental factors clearly indicates that obesity is a problem far more complex than the prevailing wisdom of 'low willpower'. Indeed, we are yet to fully understand the sophisticated interactions between genetics, physiology and cognitive behaviour that regulate energy and body weight.

Several lines of evidence suggest that our bodies have 'switches' that influence how quickly we age. These switches are not set in stone; they are adjustable and provide the potential to extend our years of youthful vigour and simultaneously postpone the troublesome conditions expressed at later ages. Diet and body weight hold the key to many of the switches and are a major factor in turning on or off components of cell ageing. 'Let food be thy medicine and medicine be thy food.' This oft-quoted adage from Hippocrates from over 2,000 years ago remains relevant today as there is a growing renaissance in our appreciation of the importance of diet in maintaining health of body and brain.

The respective diets in the Blue Zones are a good starting point for examining foods that enhance ageing. Centenarians in the Blue Zones have well-described dietary patterns that contribute to longevity and compressed ill health in later years. Exploring these diets shines a light on potential 'good foods'. Indeed, the centenarians' diet shares many features with the well-known Mediterranean diet. In brief, they are 95 per cent plants, high in fish, very low in red meat, moderately low in dairy and eggs, very low in sugars and devoid of processed foods. The Okinawa diet includes a lot of turmeric and ginger. Blue Zone residents eat a wide variety of vegetables in addition to pulses such as beans, lentils, peas and chickpeas. The diets are high in varied fruits, whole grains, nuts and seeds. At least half a cup of cooked beans and two ounces of nuts are consumed daily.

In most of the Blue Zones, cow's milk products are not included in significant amounts. People in Ikaria and Sardinia consume goat's and sheep's milk products. Eggs are consumed two to four times per week, usually one at a time and incorporated into a dish, rather than as a main protein source. In most Blue Zones, people eat up to three servings of fish each week. These are typically middle-of-the-food-chain species, like sardines, anchovies and cod, that are not exposed to high levels of mercury or other harmful chemicals. Meat is consumed sparingly. It is eaten on average five times per month, in portions that are about 60 grams or less. Rather than occupying the centre of the plate, meat is a small side; it's thought of as a celebratory food or a way to flavour primarily plant-based dishes. People in Blue Zones consume a fifth of the added sugar intake per day of North Americans. Sugar is enjoyed intentionally as a treat and not hidden in processed foods or consumed out of habit. Meals are primarily home-cooked, with breakfast as the largest and dinner the smallest. With very few exceptions, there are just four beverages: water, coffee, tea and wine. In all Blue Zones, tea is sipped daily.

In Okinawa, green tea not only provides essential antioxidants but also a healthy catalyst for socialising with family and friends. Green tea is incorporated into a meal or into a ritual for receiving visitors. It has been shown to contain catechin, which, in mice, slows down brain ageing and increases nerve circuits and the adaptability of brain nerve cells by manipulating genes which have these functions. In most of the Blue Zones, one to three small glasses of red wine are consumed per day. In Sardinia, wine consumption is part of a daily 'happy hour' ritual combining social engagement, chit chat and a couple of glasses of wine.

The Mediterranean diet is based on traditional foods that people – who were noted to be exceptionally healthy and longer living compared to Americans – consumed up until 30 years ago in countries like Italy, Greece and Spain. A recent review paper summarised information on the diet taken from a series of studies encompassing a total of 13 million participants and it was good news all the way for the Mediterranean diet. The review confirmed a robust link between the diet and reduced risk of death, cardiovascular diseases including heart attacks, some cancers, diabetes and brain diseases such as dementia. The foodstuffs included in the diet from the original studies have been expanded and the diet now broadly refers to foods detailed below with avoidance of sugars, starches and processed or refined foods.

THE MEDITERRANEAN DIET

Vegetables:	Tomatoes, broccoli, kale, spinach, onions, cauliflower, carrots, brussels sprouts, cucumbers
Fruits:	Apples, bananas, oranges, pears, strawberries, grapes, dates, figs, melons, peaches
Nuts and seeds:	Almonds, walnuts, macadamia nuts, hazelnuts, cashews, sunflower seeds, pumpkin seeds

Legumes:	Beans, peas, lentils, pulses, peanuts, chickpeas
Tubers:	Potatoes, sweet potatoes, turnips, yams
Whole Grains:	Whole oats, brown rice, rye, barley, corn, buckwheat, whole wheat, wholegrain bread and pasta
Fish and Seafood:	Salmon, sardines, trout, tuna, mackerel, shrimp, oysters, clams, crab, mussels
Poultry:	Chicken, duck, turkey
Eggs:	Chicken, quail and duck eggs
Dairy:	Cheese, yogurt, Greek yogurt
Herbs and spices:	Garlic, basil, mint, rosemary, sage, nutmeg, cinnamon, pepper
Healthy fats:	Extra virgin olive oil, olives, avocados and avocado oil

The basics of the diet are similar to Blue Zone diets. The Mediterranean lifestyle also involves sharing meals with other people and intergenerational social engagement – grandkids, parents, grandparents regularly eating together. Because the contribution of social engagement and pleasure is difficult to unravel from dietary constituents, all are recommended.

◆

Caloric restriction holds huge promise for deceleration of ageing and assisting with age-related changes in the basal metabolic rate.

We have known for some time that calorie restriction prolongs lifespan. This is so for a number of species – for example, mice, worms, fish and monkeys. In rhesus monkeys, after 20 years of reduced calorie intake, eating less than half of the monkeys' normal intake, the fasting monkeys are much younger looking, with more hair, no sunken eyes, fuller cheeks, more youthful posture and more energy than monkeys of the same chronological age who have eaten a normal diet for 20 years. Remarkably, the fasting monkeys also live 30 per cent longer.

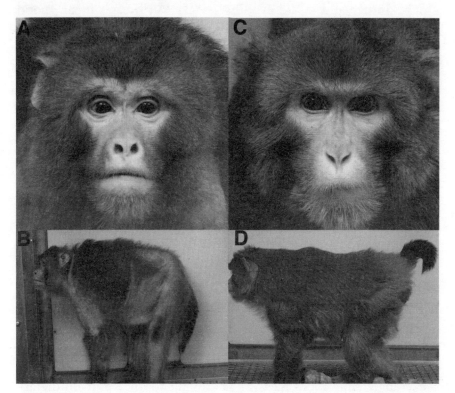

Photos A and B show a 20-year-old Rhesus monkey with normal lifelong dietary intake. Photos C and D show a monkey of the same age with 20 years of reduced calorie intake.

Ketones are chemicals that break down fat. The body uses them for energy during times of fasting and exercise. The benefits of calorie restriction and fasting is all about ketone generation.

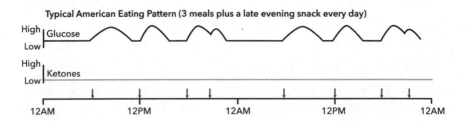

The above graph is an example of the typical eating pattern in most industrialised countries. Every day, the person eats breakfast, lunch, dinner and a late evening snack. With each meal, blood sugar levels rise and then return towards baseline over a period of several hours. Sugar is stored as glycogen in the liver. We use glycogen and therefore sugar as our main energy source when we have plenty of both. But elevated sugar levels are not good for us. Ketones are only formed when we fast and levels remain low when liver glycogen stores are plentiful. When liver glycogen levels drop, we are programmed to switch to a different means of energy production, using fatty acids to produce ketones and energy instead of glycogen. These ketones and their associated metabolic pathways are good for cells and overall health.

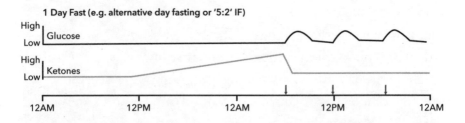

The above graph is an example of fasting for one day, followed by a three-meal feeding day (intermittent fasting). During the fasting day, glucose levels remain in the low normal range and ketone levels rise progressively, then fall when the first meal is consumed on the second day.

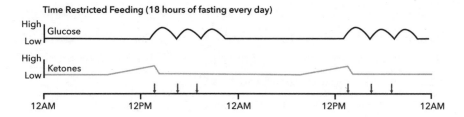

Time Restricted Feeding (18 hours of fasting every day)

The graph above is an example of an eating pattern in which all food is consumed within a six-hour time window each day. Glucose levels are elevated during and for several hours after the six-hour period of food consumption and then remain low for the subsequent 18 hours, until food is consumed the next day. Ketones are elevated during the last six to eight hours of the fasting period.

I delivered a lecture on this topic to an audience of medical doctors and one retired professor of 'Obs and Gobs' (Obstetrics and Gynaecology) was very agitated about the concept of fasting. He challenged the data, arguing that it couldn't possibly be good to have ketones. He cautioned that this was something he always tried to avoid in his patients, particularly sick diabetic patients. Of course, he was partially correct. Ketones produced because of illness are an indication of how ill someone is and are different to the ketones we strive to produce with wilful fasting. I'm glad to say that I since heard that the same Obs and Gobs professor is now an ardent disciple of caloric restriction and thriving in his mid-eighties.

There are a number of fasting programmes. For example, a 16–48-hour fast with little or no food intake and intervening periods of normal food intake, on a recurring basis. Or intermittent fasting, such as 60 per cent restriction on two days per week or every other day. Or periodic fasting, for example, a 5-day diet providing 750–1,100 kcal per day. A popular restrictive diet and my preferred choice, which many find easy to comply with, is the 18-hour fast. This is the time-restricted diet illustrated above, with food intake confined to the remaining six-hour window. So I will skip breakfast, eat two meals between midday and the evening (within six hours) and fast overnight and next morning (making up 18 hours). I find this to be the easier restriction regime to adhere to. To our knowledge, no regime is superior in respect of biological ageing, so you should choose whichever you find easiest to comply with. But whatever you choose, it will trigger a metabolic switch from glucose-based to ketone-based energy production. This in turn triggers a cascade of beneficial chemical reactions for cell preservation. All of these diets work because intermittent ketone-based energy slows cell ageing through beneficial chemical reactions.

Fasting is not for everyone, for example people with diabetes or who have a tendency to faint or feel weak, have eating disorders or are pregnant or breast feeding. If you find fasting difficult, try to curtail eating to within an 8–10-hour window and, if possible, avoid snacking. If you need a snack, eat a piece of fruit or some nuts. I found I acclimatised to fasting – even though I do a busy clinic on a number of mornings, but it did take time to get used to it. Water should be taken throughout the day – it is really important not to become dehydrated. Calorie restriction by 30-40 percent on a daily basis is also a good approach. An interesting study in obese humans with early (pre) diabetes showed that eating between 7am and 3pm and fasting for the latter part of the day and through the night significantly lowered insulin (which is a good thing for reducing fats in cells). Personally, I found this fasting regime more difficult than morning fasting – so try out a few intermittent fasting regimens until you get the one that suits you best.

You may well ask why we have evolved such that fasting confers benefits, and how it slows down ageing and diseases at a cell level. The survival and reproductive success of all organisms depends upon their ability to obtain food. We are what we eat.

Accordingly, we have evolved behavioural and physiological adaptations to survive periods of food scarcity or absence. Some organisms become dormant during periods of food scarcity; for example, yeast enter a stationary phase and ground squirrels and bears hibernate. Mammals have organs such as the liver and fat tissue in which to store energy; this enables us to fast or starve for long periods of time, depending upon the species.

In mammals, many of the health benefits of intermittent fasting are not simply the result of reduced free radical production or weight loss. Intermittent fasting also triggers responses that suppress inflammation. During fasting, cells activate pathways that enhance defences against inflammation and stress and remove or repair damaged molecules, all linked with cell ageing. Caloric restriction triggers release from fat cells of the protein adiponectin which helps to guard against heart disease and high blood pressure because of its anti-atherogenic and anti-inflammatory effects. In animals, caloric restriction reduces the chances of getting cancer and it is very likely that this is the case for humans.

Generation of ketones and reduction of sugar peaks are central to these benefits. Specifically, in humans, calorie restriction improves age-related sensitivity to insulin. In a multi-centre UK study carried out in 2017, led by Roy Taylor's group in Newcastle, diabetic patients were randomly allocated to usual care or intense caloric restriction (800 kcal per day). After one year, half of diabetics who adhered to the diet no longer required diabetic drugs, emphasising the impact of caloric restriction on remission of type 2 diabetes and insulin sensitivity.

An excellent review in the *New England Journal of Medicine* summarises the current science and concludes that fasting is evolutionarily embedded without our physiology, triggering several essential cellular functions. The authors affirmed that changing from a fed to fasting state not only helps to burn calories and lose weight but improves metabolism, lowers blood sugars, lessens inflammation and clears toxins and damaged cells – all of which improves a range of health issues from arthritis to asthma and cancer.

The question is whether we should be fasting throughout our lives to access these beneficial effects or whether later introduction of fasting can make a difference to health-related outcomes and ageing. Well, the news is also good on this front. In animals, fasting introduced at any stage in adult life shows all of the cellular benefits detailed above, even in very old animals. Humans gain benefits from fasting, no matter what age it is introduced in adult life, but the earlier one starts, the longer and more dramatic the outcomes. Try it: I've only been doing it for three years and I can honestly say I now enjoy it and I strongly encourage my patients to consider intermittent fasts.

Whereas increasing lifespan and improving health in old age is enormously attractive to most people, a lifetime commitment to a reduced calorie diet is unlikely to be adopted at a population level. As a result, caloric restriction 'mimics' are being sought – in other words, drugs or supplements which have the same effects on cells as fasting. Several have already been identified, such as resveratrol, quercetin, fisetin, metformin and rapamycin.

Resveratrol is part of a group of compounds called polyphenols, which act like antioxidants. Resveratrol extends lifespan in several species and occurs naturally in various plants, including red grapes, peanuts, prunes, blueberries and cranberries. You will probably be most familiar with references to resveratrol in red wine – which comes from the skin of grapes. Several laboratory studies have demonstrated beneficial immune protection via the action of resveratrol on the SIRT1 gene in animals and human cells. This gene is believed to protect the body against the effects of obesity and some of the diseases of ageing. So far, studies have not discovered any severe side effects, even when resveratrol is taken in large doses – though patients who are also taking blood thinners or clot reducing medications are advised to take care with intake. The dosages in most resveratrol supplements are typically lower than the amounts that have been seen to be beneficial in research. To get the dose used in research studies, consumption of up to 2,000mg of resveratrol a day is recommended. Red wine has only 5mg to 15mg in one litre. So, although wine has benefits, I am not recommending that you drink the equivalent of 2,000mg resveratrol in red wine per day – it's best to source it from elsewhere!

Quertin is another polyphenol found in fruit, particularly strawberries, nuts and herbs. It has anti-inflammatory properties, some antihistamine properties (anti-allergies) and increases anti-oxidant protection.

A new kid on the block as a fasting mimic is fisetin, which manipulates mTOR. This protein instructs insulin pathways and maintains good function of liver, muscle, white and brown fat tissue, and the brain. It appears to be very important for cell ageing. mTOR malfunctions in diabetes, obesity, depression, certain cancers and ageing cells. The relative proportions of fisetin in fruits and vegetables are strawberries (160), apples (27), persimmons (11), lotus root (6), onions (5), grapes (4) and kiwi (2). In other words, strawberries have eighty times more fisetin than kiwis. But human research on fisetin as a supplement is still in the early stages.

Rapamycin is another mTOR inhibitor and a good candidate to mimic the effects of caloric restriction. This drug may have benefits for immunity in old age as well as broader positive health effects. It is already used to complement chemotherapy drug treatments in cancer patients but clinical trials proving efficiency and safety for ageing are as yet lacking.

The action of metformin, a treatment for type 2 diabetes, also mimics caloric restriction and extends lifespan and health span in several species including rodents. Death rates are lower in diabetics taking metformin than those on other diabetic medications. This has generated a lot of interest in the drug's potential for slowing ageing. With regard to immunity, recent clinical studies have reported an anti-inflammatory role for metformin and a beneficial effect for pathways involved in mouse models of arthritis.

What can we conclude from what we have learned so far about these calorie restricting agents from early trials? Some of them are part of a healthy diet and taken as supplements may be beneficial and are unlikely to do harm. Others, such as rapamycin and metformin, need more clinical trials to establish benefits but are looking very promising – watch this space.

◆

If you want to live to be a happy 100-year-old, it's worth looking to Japan, which is the country with the longest average life expectancy in the world, at an average of 87.3 years for women and 81.3 years for men. The average lifespan of the Japanese is the highest it has ever been and is increasing. In 2019, the number of Japanese aged 90 reached 2.31 million, which included over 71,000 centenarians. So, have the Japanese found the fountain of youth? Let's delve into Japan's dietary secrets for good health and longevity.

The diet is generally lean and balanced. Staple foodstuffs include omega-rich fish, rice, whole grains, tofu, soy, miso, seaweed and vegetables, all of which are low in saturated fats and sugars and rich in vitamins and minerals that reduce the risk of cancers and heart disease. This healthy diet has led to an impressively low obesity rate, while other countries are increasingly struggling. Just 4.3 per cent of the Japanese population is obese, compared to 27.8 per cent in the United Kingdom and a staggering 36.2 per cent in the United States. Obesity is a major cause of killer diseases like diabetes, cancer and heart disease, so it goes without saying that one reason that the Japanese live longer is likely to be their diet.

The science backs it up. According to a study published in *British Medical Journal*, those who followed the Japanese government's recommended dietary regime had a 15 per cent lower death rate than those who didn't. And they start young. Japanese schools follow healthy dietary guidelines, with lunches of plenty of fruits and vegetables and very little refined sugar. Learning how to eat a balanced diet from a young age sets children up for good health for the rest of their (likely long!) lives.

Something else that is taught from youth is a Confucian teaching: *hara hachi bun me*, which roughly translates to 'eat until you are eight parts full' – similar to the Blue Zones practice. It usually takes at least 20 minutes for the brain to recognise that we are full. Smaller portions and slower eating contribute to the long lifespan of the Japanese. At mealtimes, they serve the food onto lots of small plates, sitting on the floor and eating together. Along with using chopsticks, this slows down the whole eating process, which aids digestion.

The Japanese have been drinking matcha green tea for centuries – the practice of traditional Japanese tea ceremonies that dates back over 1,000 years emphasises the important role it plays in their culture. This ancient drink is rich in antioxidants that boost the immune system and help prevent cancer, plus it even helps to preserve cell membranes, all of which combines to slow cell ageing. It also helps with digestion, energy levels and regulation of blood pressure. The secret to matcha tea's powers is in the production process. The young leaves are deprived of sunlight when growing, which increases their chlorophyll and antioxidant content. The Japanese drink this tea several times a day. Next time you reach for a cup of coffee, why not try green tea instead?

Aside from diet, other characteristics may also contribute to Japanese longevity. Around 98 per cent of Japanese children walk or cycle to school, while national radio stations broadcast 'gymnastics' each morning. The daily commute is also active, with most people walking or cycling to the train station, standing on the train, then walking to work. And it's not that they don't sit down – they just do it in a healthier way. People often sit on the floor for meals or socialising in a kneeling position known as *seiza*. The position involves resting on your shins and tucking your feet underneath your bottom. This helps to maintain strength and flexibility. Even going to the bathroom incorporates movement in Japan. Traditional Japanese toilets involve squatting, which is healthier for the bowels and muscles!

Daily physical activity continues well into the latter years of Japanese people. You'll see many older people in Japan walking or cycling. The long lifespan of the Japanese can also be put down to their excellent healthcare. Japan's healthcare system is one of the best in the world (ranked fourth by Bloomberg Efficient Health Care). Since the 1960s, the government has paid 70 per cent of all health costs and up to 90 per cent for low-income citizens. They also have advanced medical knowledge and equipment, making Japan the ideal place to get old.

It's traditional for people to care for older family members rather than admission to care homes. The psychological benefits of living with your family in old age mean that people are happier and live longer. The Japanese may also have a genetic advantage. Two specific longevity genes are more common in Japan. DNA 5178 and the ND2-237 Leu/Met genotype both play a potential role in extending life by preventing the onset of some diseases. DNA 5178 helps people resist adult-onset type 2 diabetes, strokes and heart attacks. ND2-237 Leu/Met genotype provides resistance to strokes and heart attacks. Whereas there are lots of good lessons regarding not only diet but also lifestyle to be gleaned from our Japanese friends, genetic factors may also play a role.

The Okinawans have a saying: 'Eat something from the land and something from the sea every day.' Fish is packed with nutrients that many people lack, including high-quality protein, iodine and various vitamins and minerals. Fatty species are considered the healthiest because fatty fish, including salmon, trout, sardines, tuna and mackerel, are higher in fat-based nutrients. This includes vitamin D and omega-3 fatty acids.

Omega-3 fatty acids are crucial for optimal body and brain function and strongly linked to a reduced risk of many diseases. To meet omega-3 requirements, eating fatty fish at least twice a week is recommended. If you are a vegan, opt for omega-3 supplements made from microalgae. Heart attacks and strokes are the two most common causes of premature death in the world and fish is considered one of the most heart-healthy foods you can eat. Unsurprisingly, many big studies show that people who eat fish regularly have a lower risk of heart attacks, strokes and death from heart disease. In one large UK study of 40,000 people, followed up over 18 years, fish eaters were 13 per cent and vegetarians 22 per cent less likely to have a heart attack than meat eaters. Fish also benefits the immune system and omega-3 fats in fish are especially important for the brain and the eye.

Some fish are high in mercury, so fish low in mercury such as salmon, sardines, and trout are best. High levels of mercury have been linked to cardiovascular and brain disease, including possibly dementia, but the data are not definitive and levels are rarely high enough to cause concern in adults. In general, farmed and wild fish have similar mercury content but farmed salmon contains slightly more omega-3, much more omega-6 and more saturated fat. It also has 46 per cent more calories, mostly from fat. Conversely, wild salmon is higher in minerals, including potassium, zinc and iron and in vitamin D. As evidence that fish is good for the brain, people who eat fish regularly have more grey matter in the brain centres that control memory and emotion and better memory test results.

Many of us have experienced depression at some stage in life, characterised by low mood, sadness, decreased energy and loss of interest in life. Although it isn't discussed nearly as much as heart disease or obesity, depression is currently one of the world's biggest health problems. Regular fish eaters are less likely to become depressed. Trials also reveal that, in patients who have been diagnosed with depression, omega-3 fatty acids and fish reduce symptoms and significantly increase the effectiveness of antidepressant medications, and even reduce suicidal thoughts and self-harm. In patients who had self-harmed, those randomised to omega oil supplement for 12 weeks in addition to standard psychiatric care achieved substantial reductions in markers of suicidal behaviour and improvements in overall well-being compared to placebo and standard psychiatric care. Fish also benefits our sleep. In a study of middle-aged men, a meal with salmon three times per week over a six-month period led to improvements in both nocturnal sleep and energy by day.

Red meat is a constant source of confusion for those looking to eat healthily – is red meat a good or bad food? Health benefits or otherwise of red meat consumption remain contentious. Blue Zone, Mediterranean and Japanese diets are all low in red meat, which is more common in affluent societies. One recent large-scale evaluation of combined evidence from a number of research studies examined the effects of red meat on a variety of health issues. The authors concluded that whereas there is some evidence that red meat consumption might be harmful, it is not strong enough to justify recommending that people change their dietary habits and desist from eating red meat. I have no doubt this debate will continue to rage because there are many vested interests and the outcomes are still unclear. Yet long-lived societies eat little or no red meat!

—◆—

Did you know that vitamin D is a hormone? It's the only vitamin classed as a hormone, which explains its widespread influence on so many body functions. Vitamin D was discovered in 1920. You may well have seen old photographs of young children with grossly deformed legs due to rickets, caused by vitamin D deficiency in early childhood when bones are forming. Since discovery of the cause of rickets, foods for infants have been fortified with vitamin D and rickets has virtually disappeared from Western countries but deficiency still remains an issue in adulthood and older age and in some other susceptible groups such as the obese, the immunocompromised, people who cover up in the sun and people with bowel conditions such as inflammatory bowel disease or those with dark skin. These groups should take vitamin D supplements. In Ireland, 29 per cent of 18 to 30-year-olds and one in five people over 50 are vitamin D deficient in winter and spring. One in eight people over 50 are deficient all year round and half of people over 85 are deficient. The figures are the same in the UK and other countries at high latitude where foods are not fortified. There are three sources of vitamin D: sunlight, foods and supplements. It is very hard to get enough Vitamin D from food alone if we live at high altitudes, hence supplements are necessary. Foods high in vitamin D include fatty fish such as salmon, tuna and mackerel. Beef liver, cheese and egg yolks also provide small amounts.

Vitamin D is best known for maintaining strong bones. It does so by helping the body absorb calcium from food in the gut. Calcium is one of bone's main building blocks and is needed to prevent bone thinning (osteoporosis). Osteoporosis becomes much more common as we get older, particularly in women, although men can also be susceptible to it: one in seven people with osteoporosis is male. Good diet and exercise lower the risk of osteoporosis and bone scans should be conducted at least every five years after the age of 50 to check for osteoporosis because it can be treated. If it isn't treated, bones start to fracture and it's not uncommon for someone who's experienced a fracture not to regain their previous level of function. It's so frustrating when I encounter – which happens all too frequently – a patient with osteoporotic fractures which could have been prevented by earlier treatment.

Vitamin D is important to the body in many other ways. Muscles need it for strength; nerves need it to carry messages from the brain and the immune system needs it to fight infections, including Covid-19. Our research supports a role for vitamin D in reducing the severity of Covid infections, including reducing deaths. Vitamin D may also be of benefit in age-related inflammation.

The amount of vitamin D we need each day depends on age. For preventing the most serious effects of Covid-19, our research showed that intake of at least 800IU was associated with reduced severity in the response to infection, including much less frequent admission to intensive care. Daily doses of vitamin D up to 4000IU are safe. I personally take 1000IU per day and some of my colleagues take higher doses.

◆

Before discussing antioxidants, we should remind ourselves of what they do. Free radicals are the toxic molecules formed naturally in the cell during energy production. They cause 'oxidative stress', a process that triggers cell damage and plays a role in a variety of diseases. So antioxidants are good because they mop up free radicals and prevent them from causing toxic damage to the cell and consequent diseases such as heart attack, stroke, cancer, diabetes, macular degeneration and cataracts. Examples of antioxidants include vitamins C and E, selenium and carotenoids, such as beta-carotene, lycopene, lutein and zeaxanthin.

In the USA, antioxidant supplements account for a big chunk of the total intake, i.e. 54 per cent of vitamin C and 64 per cent of vitamin E. This is where the controversy starts. In laboratory experiments, antioxidants very effectively counteract the effects of free radicals. However, supplements do not have the same beneficial effects on health in humans unless taken as part of a good diet, such as the Mediterranean diet, which naturally contains many antioxidants. The question is, why is natural dietary source so much better than antioxidant supplements?

In one study, which included almost 40,000 healthy women 45 years of age and older, vitamin E supplements did not reduce the risk of heart attack, stroke, cancer, macular degeneration or cataracts. Another large study found no benefit for vitamin C, vitamin E or beta-carotene supplements on heart disease, stroke or diabetes. The Physicians' Health Study II, which included more than 14,000 male doctors aged 50 or older, found that neither vitamin E nor vitamin C supplements reduced the risk of heart disease, stroke, diabetes, cancer or cataracts. In fact, vitamin E supplements were associated with an increased risk of stroke caused by bleeds in the brain in this study. In a study of more than 35,000 men aged 50 or older selenium and vitamin E supplements, taken alone or together, failed to prevent prostate cancer but rather increased the risk of cancer by 17 per cent.

So, given that good diets contain antioxidants and prevent aforementioned diseases, why are antioxidant supplements not of the same benefit? Some explanations conclude that the beneficial health effects of a diet high in vegetables and fruits or other antioxidant-rich foods may actually be caused by other substances present in the same foods, other dietary factors or other lifestyle choices rather than antioxidants per se. Or it may be that the effects of large doses of antioxidants used in supplementation studies are different to the amounts of antioxidants consumed in foods. Differences in the chemical composition of antioxidants in foods versus those in supplements may also influence their effects. For example, eight chemical forms of vitamin E are present in foods. Vitamin E supplements, on the other hand, typically include only one of these forms. For some diseases, specific antioxidants might be more effective than the ones that have been tested. To prevent eye disease, antioxidants that are present in the eye, such as lutein, are possibly of benefit rather than a broad spectrum of antioxidants. Other posited reasons are that the relationship between free radicals and health is more complex than previously thought. Under some circumstances, free radicals may be beneficial rather than harmful and removing them may be undesirable. Antioxidant supplements may possibly not have been given for a long enough time to prevent chronic diseases. Another likely explanation is that the microbiome that we are about to discuss may be a main mediator in the difference between diet and supplements.

In conclusion, diets high in antioxidants have multiple health benefits but there is insufficient evidence for antioxidant supplements to replace a healthy diet. It is best, where possible, to get antioxidants from foods rather than relying solely on supplements for health benefits. This is of course not something that the market wants to hear and I doubt that things will change, given the high use of antioxidants in US despite a dearth of affirmation.

———◆———

The microbiome, the bacteria in our gut, is one of the most exciting new discoveries in recent medical history. Our body houses trillions of bacteria, viruses and fungi, known as the microbiome. While some bacteria are associated with disease, others – the 'good' bacteria – are extremely important for our immune system, heart, weight and many other aspects of health. Most of the microbes that make up our microbiome are found in a 'pocket' in the large bowel. Microbes also live on the skin and in other organs such as the vagina – they are in fact everywhere in and on the body. The relationship between our microbiome and the food we eat is complex and important and may well provide valuable information for ageing.

The story starts with the Hadza tribe in Tanzania, East Africa, a hunter-gatherer tribe, living beside Lake Eyasi, with only 1,000 of the tribe in existence today. Unlike Western civilisations, the Hadza have been eating the same diet for thousands of years. To study the gut microbiome, researchers went to live with the tribe to see how different the diet and gut microbiome of the researchers was compared with that of the tribe, postulating that the Hadza microbiome would reflect our gut some hundreds of years ago when diseases such as diabetes and heart disease were so much less common.

The Hadza live surrounded by mud, in grass huts. They hunt the same animals (antelope, wildebeests, baboons and porcupines) and eat the same plants that humans had done for 3 to 4 million years, such as honey, berries, baobao and tubers. They are nomadic and movements are dictated by availability of food. The food is raw and microbial-rich. For example, after a kill the tribe eat the stomach, which is rich in microbia, and squeeze faeces out of the colon, which is then lightly cooked. The Hadza have twice as many microbes as Westerners and do not suffer from Western diseases.

Diversity in microbiome is a good thing. In order for our microbiome to thrive, our diet should be diverse and varied, in turn ensuring that our microbiome is diverse and varied. Researchers noted that it was possible to shift the diversity of gut microbiomes within 72 hours of changing a diet. Faeces are made up of microbes, alive and dead. While living with the Hadza, researchers took samples of their own faeces daily during the stay, which were itemised and examined when they were returned to the laboratory. This showed that, as the researchers ate the Hadza diet, the microbiome became more diverse even within a few days. Wait for it – some researchers also carried out 'faecal transfers' with the tribe, transferring, using a turkey baster, the tribe members' faeces into the researcher's rectum. This showed that diversity was even more evident after the faecal transfer.

This work has contributed to a wealth of studies that have expanded the possible causal role of the microbiome, not just in diseases such as diabetes, obesity and hypertension, but also in immunity and in brain health. Unfortunately, as we slip back into a Western diet, the microbiome changes back and becomes less diverse. It appears that some of our microbiome has become extinct because of restricted diversity in our diets and researchers postulate that the 'missing microbes' may well hold the answer to some of the diseases of ageing.

When we eat, microbes are triggered to attach to the food; they start to break it down, take nutrients and energy from it and produce healthy chemicals, which in turn prevent infections, positively affect mood and suppress allergies. Because microbes are predominantly in the lower gut, fats and refined carbohydrates which are absorbed higher up in the gut don't get as far as the microbes. Microbes love polyphenols, such as peanuts and seeds, which do reach them in the lower gut. To have a healthy gut, we need diverse microbes and therefore a diverse diet to keep the microbes 'interested and stimulated'. Foods high in polyphenols are listed in the table opposite.

Foods high in fibre are also particularly good for microbial diversity and for increasing the numbers of microbia. High-fibre foods include wholegrain cereals, wholewheat pasta, wholegrain bread, oats, barley and rye, berries, pears, melon, oranges, broccoli,

Spices	Herbs	Veg	Dark Berries	Fruit
Cloves	Peppermint	Globe artichokes	Black elderberry	Apples
Star anise	Oregano	Red chicory	Low bush blueberry	Apple juice
Capers	Sage	Green chicory	Plum	Pomegranate juice
Curry powder	Rosemary	Red onion	Cherry	Peach
Ginger	Thyme	Spinach	Blackcurrant	Blood orange juice
Cumin	Basil	Broccoli	Blackberry	Lemon juice
Cinnamon	Lemon verbena	Curly endive	Strawberry	Apricot
	Parsley		Raspberry	
	Marjoram		Prune	
			Black grapes	
Beverages	Nuts	Olives	Seeds	Oils
Cocoa	Chestnuts	Black olives	Flaxseeds	Extra vigin olive oil
Green tea	Hazelnuts	Green olives	Celery seeds	Rapeseed (canpola) oil
Black tea	Pecans			
Red wine	Almonds			
	Walnuts			

carrots, sweetcorn, pulses, nuts, seeds and potatoes with skin. So there is plenty of choice – but that's the point; we need all of these to keep our gut interested, stimulated and microbiome rich.

But what has any of this to do with ageing? Plenty, if not everything! The gut microbiome in long-lived persons and centenarians is very diverse. There are specific microbiota associated with longer life that we may be able to manipulate to test whether introduction of these particular microbes can be effective if introduced into the guts of people who do not have a rich and varied microbiome. This research is ongoing. But for you and me now, the message is that long-lived fit, healthy people have very diverse microbiota.

So diet is a key contributor in shaping the composition of the gut microbiota, exemplified in the contrast between the Western and the Mediterranean diet. Both diets exert distinct effects upon the composition of gut microbiota. The Western diet – rich in fats, salt and sugar – changes gut bacteria to typify the gut microbiota of obese people. In contrast, the Mediterranean diet affects gut microbiome by causing microbiome changes which we know are linked to better mental function, memory, immunity and bone strength.

One hobby horse of mine is the ubiquitous 'emulsifier'. Emulsifiers are found in all Western processed foods, such as burgers, ketchup and mayonnaise. Emulsifiers are purported to be 'safe' but they do increase the levels of microbes that produce chemicals associated with obesity and diabetes. Likewise, artificial sweeteners, although 'safe', also produce toxic chemicals via microbes. However, the doses used in laboratory experiments for both are often higher than in foodstuffs and further research on this topic is ongoing. Nonetheless, none of the diets that prolong healthy life – Mediterranean, Japanese or Blue Zone – contain refined or processed food emulsifiers.

The Mediterranean diet is full of polyphenols and dietary fibres. It is not entirely clear whether its beneficial consequences on health are due to microbiome change or to other factors associated with the diet, or a combination of these factors, but the more we stick to a Mediterranean diet, the higher the levels of good bacteria in the gut that we know are linked to successful ageing will be. Many researchers now claim that the microbiome is the missing link in understanding the relationship between the gut and food. Whatever the association, it is never too late to start a healthy diet. Changes in the microbiome occur very rapidly – even within 72 hours – and this is so for all ages. The circumstantial evidence for benefits from changing our microbiome is very strong and we have lots of choice in the foods that do this. So there's no excuse!

To keep our microbiome in good shape, two options are currently recommended in addition to dietary change: prebiotics and probiotics. Prebiotics are substances such as inulin, a water soluble fibre sourced from chicory root, on which microbes thrive. Probiotics are microbes themselves, such as the lactobacillus and bifidobacterium species. While both prebiotics and probiotics can be taken as supplements, whether you should shell out for them is another matter: there is little evidence for which prebiotics or probiotics people should consume and when it comes to probiotics, it isn't certain that the microbes will colonise your gut when they get there or if they will offer benefits to people who already have a healthy microbiome.

As we get older, we get more infections – particularly of the chest and kidney – and use more antibiotics. These reduce bacteria and microbiome in the gut. If you are taking antibiotics or have irritable bowel syndrome, there is evidence that probiotics help. Ideally, we should try to combine a prebiotic and a probiotic. Sauerkraut (finely cut and fermented raw cabbage) or kimchi (spicy fermented cabbage) are examples of foods which combine pre and probiotic properties. There is a lot of active research in this area and no doubt in the near future it will be possible to have our individual gut microbiome analysed with tailored recommendations for dietary changes based on individual patterns.

I want to share a short aside which is of interest. Remember the faecal transplant with a turkey baster in Tanzania? Well, believe it or not, this is not so out of the ordinary. Faecal transplant is a well-recognised treatment that works by giving new bacteria and microbiome to a diseased gut. This treatment is widely used in medical practice and, although the method is a bit more refined than the turkey baster, the principles are the same. Faeces from a healthy person are given via an enema to treat serious types of diarrhoea called membranous colitis, which can occur when older patients receive antibiotics.

When I was a junior doctor this was a common and dreaded complication of antibiotic use. The gut became sterile from antibiotics and devoid of microbiota, and consequently was colonised instead by a very toxic bacterium – Clostridium difficile – which took over the gut, covering the gut wall with a membrane that prevented absorption and caused severe diarrhoea – and often death. Clostridium difficile was referred to as the 'superbug'. Then along came faecal transplants with dramatic curative results. In 1958, Ben Eiseman, a surgeon from Colorado, published a paper with his team describing the successful treatment by rectal faecal transplant of four critically ill people. It took another 30 years before faecal transplants were widely applied to patients with severe antibiotic-induced diarrhoea and even longer before we recognised the gut microbiome and its curative role in faecal transplants. Remarkably, 95 per cent of patients infected with the so called 'superbug' are now cured because the transplanted faeces contain new viable and diverse microbiota that fight off the toxin. Today, worldwide, faeces from healthy people are collected, prepared, frozen and stored for use as treatment enemas.

Chapter 10

Sex and Intimacy

I LOVE THIS TOPIC IN THE CONTEXT OF AGEING BECAUSE IT IS so positive and when a physician takes the time to understand its import for a given patient, it is a rewarding experience. As medical students, we were taught to take detailed histories of all aspects of patients' lives. Taking a good history from a patient thereafter became the cornerstone of my medical practice. I teach my students that medicine is 90 per cent history and 10 per cent examination and technology. Taking a good history includes asking questions about sexuality and sexual problems. However, in reality, physicians rarely include these details in the routine evaluation. As a medical student, I took my instructions seriously and diligently interrogated patients about sex. I recall how older patients visibly changed, from being meek and passive to engaged and animated, when speaking about their sexuality.

Similar observations inspired Stacy Lindau, a gynaecologist in the University of Chicago who specialises in age-related sexual problems. In her landmark paper on a large series of older US adults, published in 2007, she reported that the majority of adults regard sexual activity as an important part of life. Most are engaged in spousal or other intimate relationships and, a substantial number of men and women partake in vaginal intercourse, oral sex and masturbation even into the eighth and ninth decades of life. But society and the media still struggle with sexuality in older persons; the topic is treated differently when compared with discussions on sexuality in the young and is considered taboo by many.

Sex is good for us. Just being physically close with another human increases brain levels of the 'cuddle hormone', making us feel happy and safe. Oxytocin is secreted by the posterior lobe of the pituitary gland, a pea-sized structure at the base of the brain. Oxytocin is known as the 'cuddle hormone' or the 'love hormone' because it is released when people snuggle up or bond socially. If oxytocin is administered to virgin rats, all of a sudden they begin acting like mothers, collecting all the pups and building nests. Prairie voles are monogamous mammals but if oxytocin in the vole's brain is blocked, they become uninterested in their mates. This hormone promotes extensive additional brain activity, including empathy and trust. In one study, couples who worked on art projects together, rather than alone, boosted levels of the hormone and injected more empathy into their relationship. People on oxytocin have been shown to be more willing to entrust someone with their money than those on placebo. And oxytocin not only increases monetary trust; people on the drug were 44 times more trusting with privacy and confidential information than those on placebo.

There's a common misconception that as people age, they lose interest in sex and the capacity for sexual behaviour. This is not the case: older people remain sexually active and continue to attribute importance to sex well beyond their fifties – for a substantial proportion, into their seventies, eighties and nineties.

Decline in desire with advancing age is not inevitable. Attitudes towards sex are determined by society as much as biology, and much of the biology can be managed by medications, creams and technologies. In the main, sexual activity is an essential part of intimate relationships and happiness as we age.

As evidence for this, our TILDA research reported 80 per cent of couples with an average age of 64 consider sex to be important and 60 per cent are sexually active at least weekly to twice per month. Recent data from an English study showed similar results. Older adults in England enjoy life more when they are sexually active and those who experience a decline in sexual activity report poorer well-being than those who maintain levels of sexual desire, activity and function in later life.

Although being sexually active is largely dependent on having a spouse or cohabiting partner, this is not exclusive and one in ten older persons who are unmarried or not cohabiting report a romantic or intimate partner and almost all are sexually active at an average age of 70, reinforcing the message that sexual activity and pleasure is not solely the dominion of the young. Lindau's more recent work showed that the frequency of sexual activity in older persons is similar to that of adults aged 18–59 in a US study in 1992.

Couples who are regularly sexually active and satisfied with their sex lives are more satisfied overall with life as a couple and have a more positive attitude towards ageing. Data on sexual activity and enjoyment of life consistently shows that sexually active individuals have a better quality of life and better relationships, are happier, less likely to be depressed and, in some studies, even live longer. Men and women who are sexually active have better memory and concentration. Sexual satisfaction and frequency are associated with better communication within couples and synchronicity in sexual desire and activity.

It's no secret that sex can help to produce that 'feel good' factor. This is largely because during sex, in addition to oxytocin, endorphins are released, which generate a happy or elated feeling. People who engage in sexual intercourse also have better mental health, with less depression and anxiety. Higher endorphins benefit the immune system, with all of the same advantages accrued from endorphin release during physical exercise. It is a big leap of faith to suggest that sex reduces diseases such as heart disease and cancer but nonetheless, there is accumulating circumstantial evidence that suggests this could be the case.

Masters and Johnson were the amazing pioneers of sexual studies who carried out groundbreaking observations of sexual activity and its biological consequences in the mid-1960s. Theirs was revolutionary work and it's fair to say that opinions were divided on the value of the studies at the time, which we now know were invaluable and have directed so much of our subsequent work in this field. The couple reported 11 years of physiological observations that involved 382 female volunteers, 18 to 78 years of age, and 312 male volunteers, 21 to 89 years of age. The research confirmed that sex is a physical activity which burns, on average, four calories a minute. During sexual activity, breathing progressively increases to rates as high as 40 breaths per minute. Heart rates also rise dramatically to 180 beats per minute, equivalent to that experienced at the peak of running very fast on a treadmill, and blood pressure also rises, again by dramatic levels of 80 mm Hg. To put this in context, during the day, blood pressure varies by about 20 mm Hg unless we are doing intense physical exercise. This explains why sex is an exercise and why it releases so many of the neurotransmitters or 'feel good' factors which are also released during exercise.

More recent studies have used wearable measurement technologies to determine energy expenditure during sex and shown that it compares with a 30-minute moderate intensity endurance running session on a treadmill and is slightly higher in men.

Despite the fact that one gains mental and physical health benefits from regular sexual activity, healthcare professionals and even media sources rarely give regular information on and encouragement to older people to explore sexual activity. In many cases, when it comes to older people and sex, doctors, nurses and others often put their heads in the sand and fail to talk about it. Such discussions could help to challenge norms and expectations about sexual activity and help people to live more fulfilling and healthier lives well into older age. Furthermore, most of the biological issues which complicate sex in later life are amenable to investigation and treatment and should be discussed. In studies in both US and UK, over half of sexually active people reported sexual problems in their sixties, seventies and eighties. This is all the more reason for the medical profession to discuss sex and sexual problems with patients at all possible opportunities because most are remediable.

—◆—

Could sex be good for the brain as we get older? In a study of almost 7,000 people aged between 50 and 90 years, volunteers were asked detailed questions about sexual activities in addition to completing tests of mental ability. The study also included a host of other assessments, meaning that the researchers were able to correct for all elements that affect mental health apart from sex and thereby look exclusively at the effects of sexuality on mental ability. Their paper was titled 'Sex on the Brain' and confirmed better mental abilities for planning and memory in sexually active older people. In other words, being sexually active was an independent benefit for brain health. The researchers speculated, very reasonably, that the benefits were because of release of oxytocin, dopamine and other endorphins – key brain neurotransmitters that control the transfer of messages from cell to cell. Other research in both humans and animals in the last decade underscores that frequent sexual activity might enhance performance on brain abilities – particularly memory. In addition to vaginal and oral sex, masturbation, kissing, petting and fondling are all associated with better memory function.

Sex is even good for animal brains. A 2010 study discovered a link between sexual activity and growth of new brain cells in male rats. Specifically, rats permitted to have sex daily over a two-week period demonstrated more new brain cells than rats only allowed to have sex once during the same period. Building on this, further male rat studies found that daily sexual activity was not only associated with the formation of new brain cells but also with enhanced brain function. In this case, the older rats that were exposed to daily sex grew new brain cells and performed better on memory tests. When the rats were deprived of sex the new cell formation stopped and the tests of memory deteriorated. The authors concluded that sexual activity is good for the brain provided it is repeated and persists. Of course, these experiments were not carried out on humans and whether they can be extrapolated has yet to be proven. Furthermore, we need studies on female rats' brains to understand their experience!

There are a number of possible explanations as to why engagement in sex might result in new brain cell formation and improved memory. In animal studies, penetrative sex is a form of physical activity which enhances cognition. Furthermore, the 'reward' aspect of intercourse may be a mechanism by which new brain cells form. The reward system is the ability to learn from positive experiences and understand motivation. Exposure to male pheromones both activates the female reward system and stimulates formation of new brain cells. Also, engaging in sex is associated with decreased stress and decreased depression. Stress and depression both blunt the formation of new brain cells. Finally, vaginal sex increases levels of serotonin and oxytocin, two neurotransmitters involved in stimulating new brain cell formation.

Although women are less sexually active than men at all ages and masturbate less frequently, they have less of a reduction in sexual desire, frequency and ability to become aroused than men do over time. The explanation for this is not clear but may be related to erectile dysfunction in men. Moreover, difficulties with sexual arousal, orgasm, dry vagina and pain decline among sexually active women aged 80 to 90 years. This may reflect the fact that the healthiest women and the healthiest partners survive into their eighties and nineties or that sexual activity sustains sexual ability.

The work of the influential Californian epidemiologist Elizabeth Barrett-Connor and her group first reported that during the menopause, women experience an increase in sexual desire followed by a decline in sexual desire, responsivity and frequency of sexual activities post menopause. However, sexuality does not disappear – it is just less apparent. The decline in frequency of sexual intercourse for women is due to lower levels of oestrogen and testosterone. Oestrogen is produced by the ovaries. As the ovaries begin to die, oestrogen levels drop. This results in a dry vagina, wasting of the labia, vulva and clitoris and thinness of the bladder wall, which causes pain during intercourse. Urinary infections after intercourse also become more frequent. Symptoms of cystitis and urinary tract infections include painful intercourse, pain passing urine, itching in the vulval area, more frequent urination, and sometimes incontinence of urine. Urinary incontinence and cystitis respond to antibiotic treatment, coupled with hormone replacement therapy or oestrogen vaginal pessaries. Sometimes, further medications, such as amitryptyline or pentosan polysulfate may help cystitis if other interventions are insufficient. For prevention, drinking cranberry juice reduces the risk of urinary infections. Regular sexual activity also helps to prevent symptoms because sexual intercourse improves blood circulation to the vagina, which maintains vaginal tissue.

Decline in sexual desire can impact women's self-esteem and quality of life and sometimes cause emotional distress leading to relationship problems. So, treatment that mitigates these unpleasant symptoms described above is important. People are often embarrassed to engage in discussion about sex 'after a certain age'. But don't be – physicians aren't, and will and can help.

Because women live longer, there is a shortage of single male partners in the same age group. But hope springs eternal and emanates from the Germans, where a study of single women reported attitudes to and experiences of unconventional sexual relationships in older women. For 91 women born between 1895 and 1936, 1 in 6 of the interviewees had had a relationship with a man younger than themselves, 4 per cent had had a lesbian relationship and 1 in 12 had had an affair with a married man in later life.

Most studies of sexual activity in later years traditionally focused on sexual dysfunction or disorders and treatments for same. However, this approach is thankfully changing and there is a global interest emerging in sexual activity, health and well-being. One large study in California looked at sexual activity and sexual satisfaction in 1,300 healthy women aged 40 to 100 years. These well-educated upper-middle-class women, average age 67 years, grew increasingly satisfied with their sex lives after they turned 40. The average period since menopause for women in the study was 25 years. Overall, two thirds of sexually active women were moderately or very satisfied with their sex life. Half reported sexual activity within the past month. For some, that heightened satisfaction came from having good sex; for others, it stemmed from the fact that they had less desire for sex and therefore lower expectations. Most were able to become aroused, maintain lubrication and achieve orgasm during sex, even after the age of 80. In fact, many were completely satisfied over the age of 80. A number of women who weren't sexually active still expressed satisfaction with their sex lives – indicating the role of intimacy and fondling in sexual satisfaction.

So, why do some women become more satisfied with their sex lives as they age? There are several possible explanations: older women are more sexually experienced and more at ease with sex; those older women who are not sexually active achieve sexual satisfaction through other intimacies such as touching and caressing, or some older women who have no intimate contact of any kind are perfectly happy without it. In contrast to younger people, older women were not thinking about sex, planning sex ahead of time or longing for sex during periods of the day but they did have sex that was satisfactory to them. These data suggest that if you hang in there, there's a good and satisfying sexual relationship for a lot of us up to the end.

Sexual activity in older age is more important for men than women. Eighty-five per cent of men in the UK aged 60–69 are sexually active, as are 60 per cent of those aged 70–79 and 32 per cent of those aged 80 and over. Studies in the US report similar levels of sexual activity across these age groups. Among sexually active men, sex twice a month or more frequently and regular kissing, petting or fondling were associated with greater enjoyment of life.

The principal sexual problem that men complain of is erectile dysfunction (ED). ED is the inability to get or keep an erection firm enough to have sexual intercourse. It's sometimes referred to as impotence, although this term is now used less often. Occasional ED isn't uncommon. Most men experience ED at some time in their life and it can happen at any age. One fifth of men have more problematic ED. Because ED becomes more common with age, treatments were therefore initially specifically marketed for the older man. Viagra, the best known of these, was introduced for ED over 20 years ago, grossing $1 billion per year since for its parent company, Pfizer. Latterly, ED treatments have acquired a growing market among younger men, in particular when used in combination with recreational drugs.

ED can occur because of problems at any stage of the erection process. An erection is the result of increased blood flow into the penis. Blood flow is usually stimulated by either sexual thoughts or direct contact with the penis. When a man is sexually excited, muscles in the penis relax. This allows for increased blood flow through the penile arteries, filling two chambers inside the penis. As the chambers fill with blood, the penis grows rigid. An erection ends when the muscles contract and the accumulated blood consequently flows out through the penile veins.

Many men experience ED during times of stress. It can also be a sign of emotional or relationship difficulties that may need to be addressed by a professional. Frequent ED, however, can be a sign of health problems – so the specific underlying health problem needs treatment in addition to treatment of ED. Because erections primarily involve the blood vessels, it is not surprising that the most common causes of ED in older men are conditions that block blood flow to the penis, such as hardening of arteries or diabetes. Another vascular cause may be a faulty vein that lets blood drain too quickly from the penis. Other physical disorders, as well as hormonal imbalances, may result in erectile dysfunction. These include high blood pressure, high cholesterol, obesity, neurological disorders and low testosterone levels. Many drugs can cause ED, including those used to treat high blood pressure and sleep disorders. Excessive alcohol is a not uncommon cause of ED. So, all of these potentially contributing factors should be considered when evaluating ED. Treatment will depend on the underlying cause and may require a combination of approaches and medications. A number of variants of Viagra have emerged since its release; all are helpful. Sometimes, testosterone therapy may also be effective if levels are low but this is not common.

When 75-year-old Diane Keaton admitted on a US show that she was 'sexually frustrated', the interviewer and interviewee bantered about her many past lovers and her expectations for future lovers. The world has moved on and the importance of sex for all ages is recognised and valued for women and for men. Thankfully, sexual activity well into older age is becoming less of a taboo topic.

Chapter 11

Loving Our Muscles
for Life

B ELIEVE IT OR NOT, IT WAS A DOUBLE-DECKER BUS, THE
Routemaster, that set the scene for groundbreaking research
into heart diseases. This is its story. The iconic Routemaster first
appeared on the streets of London in 1954 and has changed little
in style since the earliest versions. From its inception, each bus had
a driver and a conductor. The conductor was responsible for
walking up and down the bus and the stairs, selling and checking
tickets. The driver remained in the driver's hub, sedentary for
most of the long day.

Sudden death of middle-aged men from heart attacks in the
1950s was very high and referred to as 'the epidemic of heart disease'.
Unlike today, where routine post-mortems are uncommon, in
the 1950s it was common practice to carry out post-mortems on
all sudden deaths. Two London pathologists, Jerry Morris and
Margaret Crawford, observed that they appeared to be doing
more post-mortems on bus drivers than bus conductors and more
on desk-based post office workers than postmen. It occurred to

them that heart disease might be more frequent in people with inactive jobs. They proposed that sedentary jobs might be a cause of the heart attack epidemic, given that bus conductors and postmen were much more physically active than bus drivers and office workers. So, in order to explore their hypothesis, Morris and Crawford contacted all of the pathologists in the UK and requested details of post-mortems on all male deaths together with information on occupational histories. Almost 90 per cent of pathologists complied. This is a remarkably high response rate and demonstrates the collegial collaboration among their peers which would be unlikely on this scale today!

With this high response, they set about studying the details of post-mortem records and occupational histories of 5,000 men. This confirmed their suspicion, showing that sedentary occupations were associated with early death and that these men had died because of blockages in the arteries to the heart and therefore of heart attacks. For the first time, we had clear evidence that sedentary occupations were more likely to kill than jobs that included regular physical activity, such as walking. This observation opened up a whole new world of investigation that carries right through to today as we continue to dig deep into the biological reasons for an association between physical activity and heart disease.

Since Morris and Crawford's original work, there have been thousands of papers on the association between exercise and heart disease. For example, in one large analysis of almost a million people followed for 20 years, inactive people were 40 per cent more likely to die early than those who were regularly physically active. So, thank you Routemaster for driving us along this important new route!

Although the information on exercise and heart disease is compelling and I wax lyrical about it when I can, during one radio interview I was abruptly halted in my tracks as I embarked on my well-rehearsed narrative. The interviewer stopped me with, 'I am sick and tired of hearing about exercise and diet. People are bored with this. Personally, I don't believe it matters as much as you medics want us to think.' I understood where he was coming from. It was 'same old, same old'. I resolved after that to take a fresh approach when speaking about health behaviours. It is not enough to make a recommendation without explaining the background to justify the reasoning – or, in other words, to go back to first principles and the biological rationale for why health behaviours, including exercise, have their effects. The scrutiny starts here.

There are a number of ways in which exercise benefits the heart. It improves blood circulation, which reduces the risk of clotting in arteries. The heart is of course a muscle, like those elsewhere in the body, and regular exercise helps to keep it toned and strong. As the heart becomes stronger, the heart rate lowers because fewer beats are required to pump the same amount of blood around the body. All of this reduces pressure on the heart. A stronger heart pumps more blood with less effort. If the heart can work less to pump, the force on the arteries decreases, lowering blood pressure. The consequent lower blood pressure further benefits the heart given that high blood pressure puts unwanted backpressure on the heart muscle. Furthermore, 'good cholesterol', or HDL-cholesterol, which is 'good' because it lowers the risk of thickening of arteries, increases with exercise. Thickened arteries eventually block, which causes heart attacks.

Regular physical activity also improves mental health and well-being, prevents or alleviates depression and increases vitality and an optimistic approach. The brain recognises the start of exercise as a moment of stress and thinks that one is either fighting the enemy or fleeing from it. In response, it releases a protein called brain-derived neurotrophic factor (BDNF). BDNF is protective against stress, which partly explains why we often feel so at ease and in a happier state and issues and problems appear clearer after exercising. BDNF released during exercise supports the growth of new nerve cells which further enhance brain function, brain health, feel-good factor and cognitive performance.

As early as 1905, a publication in the cruelly named *American Journal of Insanity* described the benefits of exercise as a treatment for depression. Since this early publication, a number of chemicals released by the brain during exercise that are important both for prevention and treatment of depression and anxiety have been discovered. These include opiates, cannabinoids and endorphins, as well as BDNF. Exercise confers additional psychological benefits, like self-esteem, the sense of achievement, being in control and having a purpose. It adds variety and in some circumstances incorporates social engagement and interactions with friends. Most of us have experienced feeling too tired to move, happily slumped in front of the telly, before forcing ourself to take a walk. On return, we feel revived and invigorated. Exercise makes us feel good and this is so even if we are depressed. Yet despite the evidence for exercise being protective for depression, our TILDA research has also shown that physical activity levels are low in adults who suffer with depression. Lack of motivation which frequently accompanies depression may explain this. So, the message has to be conveyed more forcefully that even low doses of physical activity, a minimum of 150 minutes of walking per week, are protective against depression and that vigorous-intensity exercise, such as jogging, cycling, swimming and rowing, will have even more powerful benefits.

One of the most exciting discoveries in brain science has been the realisation that we can grow new nerve cells. Up until now, it was assumed that we were born with a certain number of brain cells and that as we get older, we lose cells until some of us eventually develop dementia. This is not necessarily the case. It has been long recognised that exercise sharpens certain cognitive skills. Over the past 20 years, we have begun to get at the root of how this might occur.

Remarkably, exercise increases the size of the hippocampus, which is the seat of learning and memory in the brain. Normally, the hippocampus shrinks in late adulthood – that is, the number of nerve cells declines, leading to impaired memory and ultimately increased risk for dementia. Physical activity slows down the hippocampal shrinkage. Even in older adults, studies have shown that aerobic exercise training increases the size of the hippocampus, leading to improvements in memory. Exercise training reverses the age-related loss in volume by as much as two years – nothing else has such a dramatic effect. The increase in hippocampus size also increases release of BDNF, a win-win all around. Furthermore, aerobic exercise training increases the cells in other areas of the brain involved in major cognitive tasks, including the ability to plan and to prepare complex tasks and reactions.

Cathepsin B is another new kid on the block, recently discovered to enhance brain function, which is triggered by exercise. Running in particular elevates cathepsin B, which is secreted by muscle cells and promotes and accelerates growth of new nerves. I expect that we will hear a lot more of cathepsin and exercise in the near future.

Endorphins are one of the best-known feel-good chemicals released in the brain during exercise – they also minimise the discomfort of exercise and block the feeling of pain. Both BDNF and endorphins are responsible for the feelings of euphoria with physical activity. The somewhat scary part is that they have very similar addictive physiological behaviour as morphine, heroine or nicotine. The big difference, of course, is that this addiction is really good for us.

One of the greatest fears that people express regarding getting older is of developing dementia. There is an emerging consensus that exercise in midlife prevents or delays dementia in later life. Some studies suggest that the effect is as large as a reduction of 30 per cent – in other words, people who exercise regularly are less likely by a third to get dementia. However, as yet, this is difficult to prove definitively because so many other elements linked to dementia are also influenced by exercise, such as weight, blood pressure, education, occupation and diabetes. Most of the studies that have looked at the relationship between exercise and dementia have either looked back over the amount of regular exercise that someone did throughout their life and so were dependent on a person's ability to accurately recall this or started to study people in their forties and fifties and followed them thereafter. The latter is the best type of study to really address the question properly but clearly are long term and most are still in progress.

Mice are a good model to study the link between exercise and dementia and provide a faster conclusion. A mouse lives for two to three years. As part of a research study, mice had genes modified so that they were more likely to get dementia. In these mice, exercise protected against developing dementia and BDNF was key to this prevention.

With an accumulating wealth of data on the benefits of physical activity on brain function, clinicians have begun to prescribe exercise to patients with brain diseases such as Parkinson's and Alzheimer's as well as other brain disorders, from epilepsy to anxiety. Many clinical trials of exercise interventions for age-related brain disorders are underway. Promising results could further bolster the use of exercise as a neurotherapy.

◆

You will recall that we discussed the importance of inflammation in the ageing of cells and how low or no background inflammation slows down ageing and higher background inflammation accelerates ageing. Exercise isn't just good for the heart, blood vessels and brain. Because it lowers background inflammatory states in the body, it leads to a reduction in all of the conditions that become so much more common as we get older and are linked to inflammation. These include arthritis, cancer, diabetes and strokes. I'll pause to explain inflammation and accelerated ageing in this context.

If we get an infection, we mount an inflammatory response that 'eats up' the infecting agent. This is good and it's what we want to happen. Once it has dealt with the infection, the inflammatory response recedes. However, if the inflammatory response continues to be active in the background, this is bad for cells and causes release of toxic proteins, which in turn further increase inflammation. So we only want inflammation when we have an infection or other insult to the body, otherwise it should be asleep in bed and not troubling our systems.

Background inflammation is closely linked to body fat. Fat cells produce toxic proteins which trigger inflammation. The fat that is most likely to produce these toxic proteins is the white fat that settles on our bellies and around our internal organs – this is why a big belly is bad news. Muscle mass declines and fat mass increases as we age, and these toxic proteins consequently increase and cause chronic low-grade inflammation. Regular physical activity decreases fat, including the fat most likely to produce pro-inflammatory conditions.

Fat cells also make the immune response less efficient. We saw this with Covid-19, where obesity was one of the principal risk factors for suffering from severe consequences, including death. This is dramatically illustrated by a recent French study in which the need for mechanical ventilation in the ICU in patients with Covid-19 was seven times higher for those who were obese (with body mass index (BMI) greater than 35 kg/m) compared with those with a lower BMI (greater than 25 kg/m). Two of my clinical colleagues who have been overweight or obese for as long as I have known them recognised this association very early on in the pandemic and I quite literally did not recognise them when we met for the first time after a number of months – they had deliberately lost so much weight.

Currently, understanding how to improve defences against infections is a global priority. Both viral and bacterial chest infections are less frequent in people who take regular exercise because physical activity boosts the immune system and causes the regularisation of the inflammatory responses. Physical activity makes a difference to these defences. Exercising muscle also releases enzymes called myokines that transiently block harmful inflammatory proteins and promote release of other proteins which are anti-inflammatory, providing another potent counter-attack to the chronic background inflammation that can characterise ageing.

Many people think that its 'too late to start exercising' or 'I've missed the boat' – not so. Introduction of exercise at any age changes immune responses for the better. There is strong evidence to support the fact that it is never too late to start or to increase exercise. Many studies of exercise performed one to six times per week over a period ranging from six weeks up to ten months have shown multiple positive effects on the immune system and inflammation, even in old age.

One common cause of debilitating winter infections is influenza, the 'flu'. Influenza is a viral infection that attacks the respiratory system, the nose, throat and lungs. People over 65 are more prone not only to influenza but also to severe side effects from influenza.

It is heartening to know that exercise can improve not only the body's response to flu but also responsiveness to the flu vaccine. Vaccination is recommended for all people over 60, in addition to healthcare workers of all ages and anyone who is vulnerable because of co-existent diseases. Unfortunately, the vaccine is not at all as effective in older ages as in young adults: it works in 90 per cent of young adults but only 50 per cent of people over 65. Anything that can improve this response is important and exercise helps to do just that. In one elegant study, doing aerobic exercise for the three months before having the flu vaccine significantly improved vaccination responsiveness in particular.

Despite the fact that regular physical activity is associated with these major health benefits, ageing is accompanied by a sharp decline in both duration and intensity of physical activity and the majority of adults fail to meet the World Health Organization (WHO) recommended guidelines of 150 minutes of aerobic exercise per week. The figures in Ireland and the UK are frankly shameful: almost two thirds of persons 50 years of age and over fail to meet the recommended criteria.

In a large British study, adults aged 40 and older reported spending more time on the toilet each week than walking – an average of 3 hours and 9 minutes on the toilet, compared with 1 hour and 30 minutes of walking. You may well wonder who was conducting the timing for that striking study! Furthermore, just one in ten UK adults were aware of the periods of time recommended for physical activity. Work is the biggest barrier to being more active, with 20 per cent of people citing 'being too busy with work' as a reason for not exercising. Yet productivity is better after exercise. Another pertinent problem is that two thirds of us spend at least six hours each day sitting, a factor which also significantly increases the risk of an early death.

◆

Charles Eugster, a retired dentist, delivered an inspirational TED talk on exercise and ageing at the age of 95. He described how he had started bodybuilding when he was 87 years old.' His story up to then was a familiar one. In youth, he had been a champion sprinter but he grew less and less active as he aged. The athletic glories of his youth gave way to sedentary married life. Summer days rowing and boxing slowly transitioned into evenings in front of the telly – a familiar narrative. For 40 years, Eugster put his athletic pursuits on the back burner as his children and dental practice grew. But idleness did not sit well with the British sprinting champion so he began re-sharpening his athletic edge in his mid-sixties. Eugster started skiing and pulling oars again, beginning a remarkable run in competitive sports. For two decades, he went on to dominate senior rowing, winning 36 masters gold medals.

Eugster's efforts rewarded him but despite this, he noted a deterioration of his body. At 85 years old, Eugster was widowed by his second wife and his muscles had slacked considerably. He had a 'pancake butt', as he put it, and that spurred him into a new pursuit: bodybuilding. Eugster wanted muscle and an Adonis body. He craved strength and a longer life. So he began hitting the iron at age 87. He started lifting weights, sprinting again and supplementing with whey protein. Success soon followed. He won three bodybuilding world titles and broke the 200-metre and 60-metre world records for the 95-plus age group. He travelled the world telling people of all ages about the benefits of bodybuilding, healthy eating and active living. He called on his audiences to never retire, to activate their bodies and minds, and to always pursue excellence.

From the age of 50, we lose muscle mass every year. The loss of muscle mass is aligned to loss of muscle strength, and muscle power. Protein supplementation in addition to muscle strengthening exercises is required for optimum effect. Our bodies are those of hunter-gatherers, designed for physical activity. One estimate says a modern equivalent to the typical exertions of a hunter-gatherer is walking or running 20 kilometres per day, with frequent squatting, not sitting. The hunter-gatherer had to search for food and to use his mind all the time. So, in addition to exercise, current recommendations are to stand where possible and to stand up at 45-minute intervals during prolonged sitting. This is good for 'waking up' our physiological systems and improving brain blood flow. In summary, a combination of aerobic and muscle strengthening exercises coupled with regular standing-up when sitting for prolonged periods are preferred and most closely aligned to how we have evolved.

The concept of sarcopenia is relatively new in medicine but is fast gaining momentum. It is certainly something that I see many times each day when dealing with older patients and particularly people who have been unwell for a long period of time or who have had a fall. Sarcopenia is closely aligned with physical activity and exercise. Its name comes from the Greek words *sarx* and *penia*, signifying 'loss of flesh', which is indicative of the core features of sarcopenia – skeletal muscle loss. It is a progressive and generalised ageing muscle disease characterised by loss in muscle mass, weaker muscle strength and infiltration of muscles with fat.

The main factors that cause sarcopenia are ageing, chronic diseases, low physical activity and poor nutrition. We lose 15 per cent of muscle strength due to decline in muscle mass every ten years after age 50. This loss in strength accelerates after the age of 70. This is why it becomes even more important to increase exercise, not decrease it with age and to ensure that we do both aerobic and resistant exercises. We have to work much harder after age 50 and harder again after 70 to prevent age-related sarcopenia. Studies vary regarding how common sarcopenia is but some estimates are that up to two thirds of people over 70 have it. Of course, once sarcopenia sets in, it becomes more difficult to reverse and physical activity declines further such that it becomes a vicious cycle, making it more challenging to counteract age-related weakness in skeletal muscle and offset or reverse sarcopenia. So, if you have a bad cold and are confined to bed for a few days, be aware to put effort into keeping muscles moving when in bed and building up exercise programmes once recovery starts.

What can be done to prevent or reverse sarcopenia? The solution is exercise and diet. The type of exercise is important. Whereas aerobic exercise is a must, it is not sufficient without additional resistance exercises from midlife onwards. Because loss of muscle mass is generally gradual, beginning as early as age 30 and accelerating after age 60, those who have been involved in physical activity from an early age do have an advantage. The better our starting muscle mass before decay starts, the higher the reserve capacity and the less the impact of future loss of muscle. But again, I repeat, it is never too late and we do derive benefit from resistance programmes at all ages.

Resistance exercise mitigates the effects of ageing on the nerves that feed skeletal muscles as well as on skeletal muscles themselves. A properly designed training programme will enhance muscular strength and power. At a cellular level, oxidative stress is improved and the 'energy powerhouse' of the muscle cells – the mitochondria – function more efficiently with resistance training. A programme should include an individualised, periodised approach, working towards two to three sets of one or two multi-joint exercises per major muscle group, two or three times per week. Programmes should be progressive. The earlier that one is commenced, the better, but whatever age one starts, you will feel the benefits. If training pauses or stops, there will be a regression in muscle strength and fat tissues will infiltrate the muscles. So try to keep it up and if you pause – which almost all of us do – start up again as soon as you can.

Despite the known benefits of resistance training, only 8 per cent of adults over 75 years of age in the United States participate in muscle-strengthening and resistant exercises as part of their leisure time. Reported barriers to participation include fear, health concerns, pain, fatigue, lack of social support and, of course, lack of awareness of the benefits. I do regular supervised sessions with a physical trainer. In this way, I am best motivated and he ensures progression of the resistance programmes. Wouldn't it be great if more recognition and support was given to supervised training and adults could avail more readily of affordable programmes? The longer-term benefits would outweigh subsidy costs, provided that those taking part remained true to the levels of participation required.

If you do not at present do resistance exercises to complement aerobic exercise then I recommend that you start to prevent or reduce sarcopenia. This is what Charles Eugster recognised and promoted – the value of hitting the iron, even aged 87. Studies confirm Eugster's convictions and show that even in people aged 90 and over, resistance exercises are feasible and make a difference to strength and overall well-being.

Supplements to enhance muscle power are not the exclusive dominion of the young body builder. Given that the production of protein is impaired in ageing, that proteins are key to the strength of muscles and that muscle wasting accelerates with age, including sarcopenia, protein supplements should be used to complement resistant exercise programmes. The most appropriate are supplements that target protein synthesis and thereby muscle metabolism and muscle strength, such as whey protein. One recent trial of 380 adults with sarcopenia manifesting with low muscle power and low muscle mass showed that the group who were treated with daily whey protein, which was leucine (an amino acid), and vitamin D for a three-month period had a significant improvement both in muscle mass and in strength with no side effects from the supplements. This was in the context of already having muscle wasting so the results are very promising. I take a protein whey drink after each resistance training session.

Vitamin E molecules with their antioxidant and anti-inflammatory capabilities also enhance muscle regeneration and reduce sarcopenia. Animal and human experimental studies show that vitamin E benefits new muscle formation and better muscle strength. So, for the purposes of better muscle function, vitamin D, vitamin E, omega fatty acids, amino acids – particularly leucine provide the answer and appear to work when coupled with aerobic and resistance exercise.

Exercise and diet are some of the most important modifiable factors for biological ageing and, as you now appreciate, there are a host of different exercise and healthy food options. As we get older, the inclination is to slow down with each year. I propose instead that we have a target whereby we strive to do a little *more* exercise with each passing year.

———◆———

I hope that you have enjoyed reading the book as much as I have enjoyed sharing my experience of over 35 years in clinical and research practice in this fascinating topic, and, in particular, sharing the findings from the study I established and lead, TILDA, together with other global longitudinal studies. I expect that some of you are keen to see how you operate on different performance tasks compared with chronological peers and to this end I have included some tests that cover the principal areas we have discussed. At the end of each test you will find a graph illustrating the normal distributions from the TILDA study so that you can benchmark your performance with people of similar age and sex. Happy testing!

Test Yourself

TILDA IS THE IRISH LONGITUDINAL STUDY ON AGEING. 'Longitudinal' indicates that a study involves repeatedly observing and recording the same things over a period of time to identify trends and fluctuations. TILDA has been studying the same 9,000 participants for 12 years, testing them in great detail every two years. The sample was randomly selected in a particular way at the start of the project to make it a 'representative' sample of people aged 50 years and older in Ireland. Therefore, we can generalise the findings to the entire population and consequently have generated 'normative' graphs from the data.

You now have an opportunity to try some of the tests used in TILDA to evaluate ageing and apply your test results to the population graphs to assess how you perform in comparison with those of a similar age. Although the graphs apply to people 50 and older, younger readers can also do the tests and their scores should be close to the long dashed line. In the case of happiness, you can see how they compare with older persons. If you are below average on any of the quality of life domains – that is, close to the short dashed line – then you should consider the elements we discussed in the chapters on friendship, laughter, downtime, diet, sexuality and cold water to improve your ratings. The tests that I have selected assess quality of life, perceptions of ageing, levels of worry, depression, anxiety, loneliness, purpose and how long you can stand on one leg! All of which are strong indicators of biological ageing.

QUALITY OF LIFE – CASP-12

How would you rate your quality of life? This assessment picks up on the important features that drive how much we feel we are getting out of our lives – control, autonomy, pleasure/happiness and fulfilment of potential. Higher scores on each dimension represent higher quality of life. Test each separate component and then add up the individual scores to get the total score. Compare the score for each of the components with the general population – you should be near to the long dashed line for a really good result.

This test measures the different aspects of life quality.

Put a circle around your response to each item then add the numbers together to yield your overall score for each section. Please do not leave any items blank.

Control – the ability to actively participate in one's environment

	Often	Sometimes	Not Often	Never
My age prevents me from doing the things I would like to.	0	1	2	3

I feel that what happens
to me is out of my control.

0	1	2	3

I feel free to plan for
the future.

3	2	1	0

I feel left out of things.

0	1	2	3

Total: _____

Autonomy – the right of the individual to be free from unwanted interference

	Often	Sometimes	Not Often	Never
I feel that I can please myself in what I can do.	3	2	1	0
My health stops me from doing the things I want to do.	0	1	2	3
Shortage of money stops me from doing the things that I want to do.	0	1	2	3

Total: _____

Pleasure – the sense of happiness or enjoyment derived from engaging with life

	Often	Sometimes	Not Often	Never
I look forward to each day.	3	2	1	0
I feel that my life has meaning.	3	2	1	0
I enjoy being in the company of others.	3	2	1	0

Total: _____

Self-realisation – the fulfilment of one's potential

	Often	Sometimes	Not Often	Never
I feel satisfied with the way my life has turned out.	3	2	1	0
I feel that life is full of opportunities.	3	2	1	0

Total: _____

Overall Total Score

Add up the four totals above for your overall total score, generated by summing scores for control, autonomy, self-realisation and pleasure.

Overall total score: _____

How do you compare?

Find your age along the horizontal axis and total score for each category along the vertical axis to see where you are on the scale. Nearest to the solid line is average; towards the long dashed line (95th percentile) is above and towards the short dashed line (5th percentile) below. Ninety per cent of people are between the limits indicated by the long and short dashed lines.

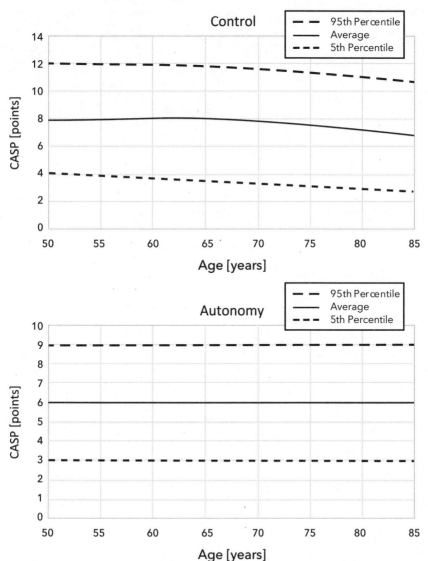

Pleasure

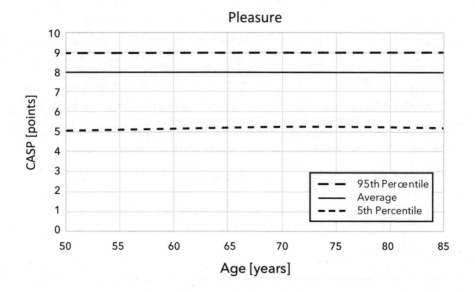

Self-Realisation

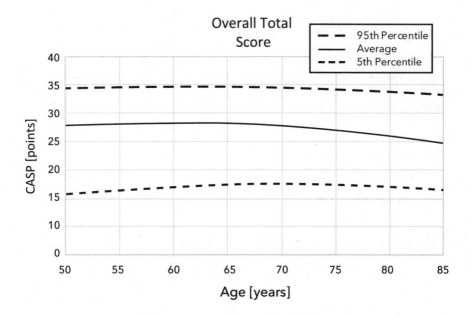

Overall Total Score

THE PENN STATE WORRY QUESTIONNAIRE (PSWQ-A)

Are you a worrier? This test measures the different dimensions of worry and anxiety. A higher score indicates greater fears or worries. If you are above the average – i.e., close to the short dashed line – then you should consider the mechanisms we discussed in Chapter 6 to reduce stress. For the purposes of this test, a lower score means that you have less fear and worry and are closer to the long dashed line.

How to score: Put a circle around your response to each item then add the numbers together to yield your overall score. Please do not leave any items blank.

	Not at all Typical		Somewhat Typical		Very Typical
My worries overwhelm me.	1	2	3	4	5

	Not at all Typical		Somewhat Typical		Very Typical
Many situations make me worry.	1	2	3	4	5
I know I should not worry about things, but I just cannot help it.	1	2	3	4	5
When I am under pressure, I worry a lot.	1	2	3	4	5
I am always worrying about something.	1	2	3	4	5
As soon as I finish one task, I start to worry about everything else I must do.	1	2	3	4	5
I have been a worrier all my life.	1	2	3	4	5
I have been worrying about things.	1	2	3	4	5

Total: _____

How do you compare?

Choose your age and total score and see where you are on the scale. Nearest to the black line is average; towards the long dashed line (95th percentile) is above and towards the short dashed line (5th percentile) below average. Ninety per cent of people are between the limits indicated by the long and short dashed lines.

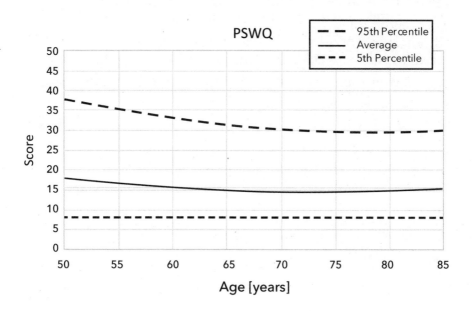

PERCEPTIONS OF AGEING

In Chapter 1 we discussed how one's perceptions of how one aged actually influenced future pace of ageing. The younger you believe yourself to be, the slower your pace of ageing. Below are a number of tests of your perception of ageing. The closer you are to the short dashed line, the better your perception of ageing. The tests measure the changes in perception (timeline) of the benefits of ageing, your control over ageing benefits, what you perceive to be the downside of ageing, and whether you consider that you have control over these 'downsides' and whether any of your negative perceptions vary through time.

How do you view ageing? This test measures the different aspects of ageing perceptions. Higher scores indicate greater agreement with the specific perception of ageing.

Put a circle around your response to each item then add the numbers together to yield your overall score for each section. Please do not leave any items blank.

Timeline Acute/Chronic – the extent to which awareness of one's ageing is constant

	Strongly Disagree	Disagree	Neither	Agree	Strongly Agree
I am conscious of getting older all of the time.	1	2	3	4	5
I am always aware of my age.	1	2	3	4	5

	Strongly Disagree	Disagree	Neither	Agree	Strongly Agree
I always classify myself as old.	1	2	3	4	5
I am always aware of the fact that I am getting older.	1	2	3	4	5
I feel my age in everything that I do.	1	2	3	4	5

Total: _____

Consequences positive – awareness of the benefits of ageing

	Strongly Disagree	Disagree	Neither	Agree	Strongly Agree
As I get older I get wiser.	1	2	3	4	5
As I get older I continue to grow as a person.	1	2	3	4	5
As I get older I appreciate things more.	1	2	3	4	5

Total: _____

Emotional representations – one's emotional responses to ageing

	Strongly Disagree	Disagree	Neither	Agree	Strongly Agree
I get depressed when I think about how ageing might affect the things that I can do.	1	2	3	4	5
I get depressed when I think about the effect that getting older might have on my social life.	1	2	3	4	5
I get depressed when I think about getting older.	1	2	3	4	5
I worry about the effects that getting older may have on my relationships with others.	1	2	3	4	5
I feel angry when I think about getting older.	1	2	3	4	5

Total: _____

Control positive – perceived control over the benefits of ageing

	Strongly Disagree	Disagree	Neither	Agree	Strongly Agree
The quality of my social life in later years depends on me.	1	2	3	4	5
The quality of my relationships with others in later life depends on me.	1	2	3	4	5
Whether I continue living life to the full depends on me.	1	2	3	4	5
As I get older there is much I can do to maintain my independence.	1	2	3	4	5
Whether getting older has positive sides to it depends on me.	1	2	3	4	5

Total: _____

Consequences negative – awareness of the downsides of ageing

	Strongly Disagree	Disagree	Neither	Agree	Strongly Agree
Getting older restricts the things that I can do.	1	2	3	4	5
Getting older makes me less independent.	1	2	3	4	5
Getting older makes everything a lot harder for me.	1	2	3	4	5
As I get older I can take part in fewer activities.	1	2	3	4	5
As I get older I do not cope as well with problems that arise.	1	2	3	4	5

Total: _____

Control negative – perceived control over negative experiences of ageing

	Strongly Disagree	Disagree	Neither	Agree	Strongly Agree
Slowing down with age is not something I can control.	1	2	3	4	5
How mobile I am in later life is not up to me.	1	2	3	4	5
I have no control over whether I lose vitality or zest for life as I age.	1	2	3	4	5
I have no control over the effects of getting older on my social life.	1	2	3	4	5

Total: _____

Timeline cyclical – the extent to which one experiences variation in their awareness of ageing

	Strongly Disagree	Disagree	Neither	Agree	Strongly Agree
I go through cycles in which my experience of ageing gets better and worse.	1	2	3	4	5
My awareness of getting older comes and goes in cycles.	1	2	3	4	5
I go through phases of feeling old.	1	2	3	4	5
My awareness of getting older changes a great deal from day to day.	1	2	3	4	5
I go through phases of viewing myself as being old.	1	2	3	4	5

Total: _____

How do you compare?

Find your age along the horizontal axis and score for each category along the vertical axis to see where you are on the scale. Nearest to the solid black line is average; towards the long dashed line (95th percentile) is above and towards the short dashed line (5th percentile) below average. Ninety per cent of people are between the limits indicated by the long and short dashed lines.

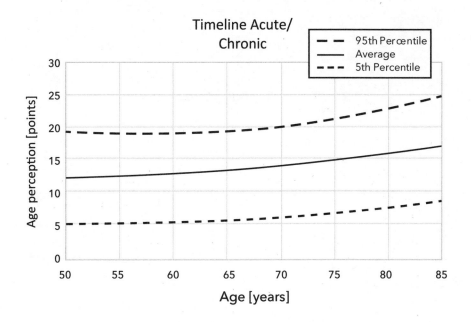

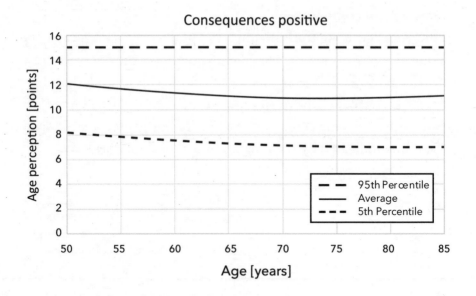

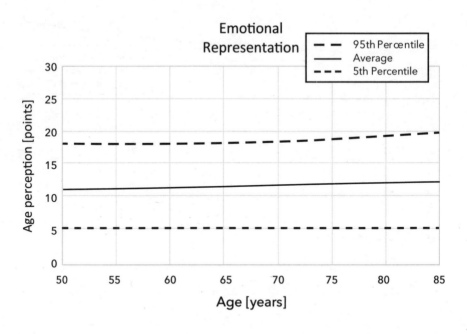

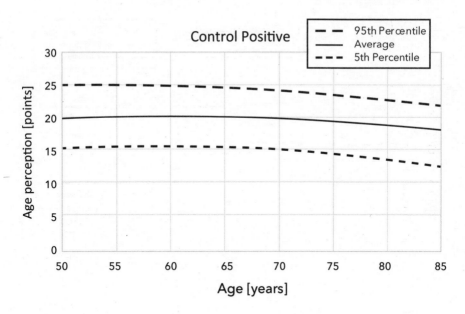

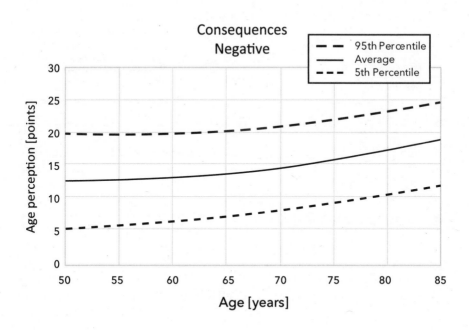

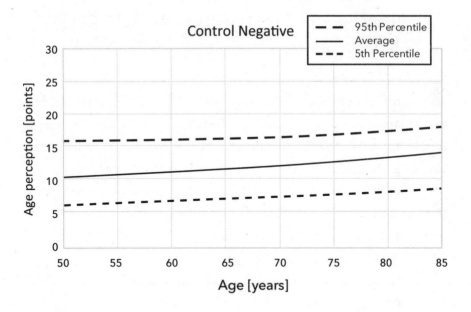

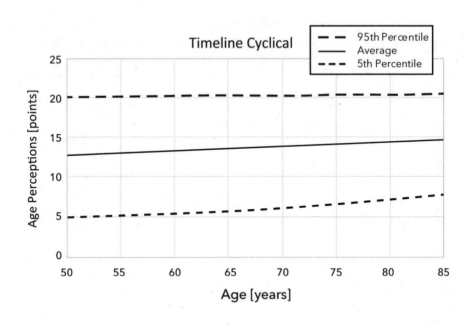

PURPOSE IN LIFE SUBSCALE OF THE RYFF PSYCHOLOGICAL WELL-BEING SCALE

Having a purpose in life is important for successful ageing. Most successful super-agers have purpose. Scientists agree that we can create purpose for each day. This can be a big task – such as employment – or small meaningful task – like household chores, assisting neighbours and friends, volunteering, gardening and other hobbies, including creativity. Grandparenting gives great reward and purpose to many. Your score should be close to the long dashed line. Once you have summed your total score, plot it against your age on the graph.

This test gives a measure of purpose in life, one of several measures of psychological well-being.

Put a circle around your response to each item then add the numbers together to yield your overall score. Please do not leave any items blank.

	Strongly Disagree	Disagree	Disagree Slightly	Agree	Slightly Agree	Strongly Agree
I enjoy making plans for the future and working to make them a reality.	1	2	3	4	5	6
My daily activities often seem trivial and unimportant to me.	6	5	4	3	1	0

| I am an active person in carrying out the plans I set for myself. | 1 | 2 | 3 | 4 | 5 | 6 |

| I don't have a good sense of what it is I'm trying to accomplish in life. | 6 | 5 | 4 | 3 | 2 | 1 |

| I sometimes feel as if I've done all there is to do in life. | 6 | 5 | 4 | 3 | 1 | 1 |

| I live life one day at a time and don't really think about the future. | 6 | 5 | 4 | 3 | 1 | 1 |

| I have a sense of direction and purpose in my life. | 1 | 2 | 3 | 4 | 5 | 6 |

Total: _____

How do you compare?

Find your age along the horizontal axis and your total score for the questions above along the vertical axis to see where you are on the scale. Nearest to the solid black line is average; towards the long dashed line (95th percentile) is above and towards the short dashed line (5th percentile) below average. Ninety per cent of people are between the limits indicated by the short and long dashed lines.

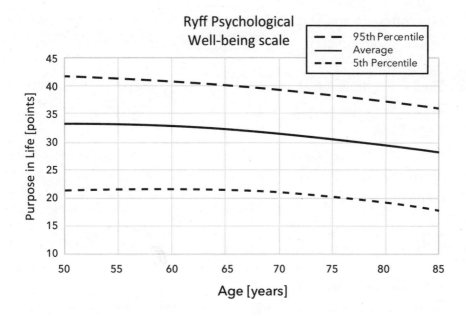

Ryff Psychological Well-being scale

- - - 95th Percentile
—— Average
- - - 5th Percentile

UCLA LONELINESS SCALE

This test is a measure of loneliness. Higher scores represent greater feelings of loneliness.

The questions are about how you feel about different aspects of your life. For each one, please say how often you feel that way.

Put a circle around your response to each item then add the numbers together to yield your overall score for each section. Please do not leave any items blank.

	Often	Some of the time	Hardly ever or never
How often do you feel you lack companionship?	2	1	0
How often do you feel left out?	2	1	0

How often do you feel isolated from others?

2   1   0

How often do you feel in tune with the people around you?

0   1   2

How often do you feel lonely?

2   1   0

Total: _____

How do you compare?

Find your age along the horizontal axis and total score along the vertical axis to see where you are on the scale. Nearest to the solid black line is average; towards the long dashed line (95th percentile) is above and towards the short dashed line (5th percentile) below average. Ninety percent of people are between the limits indicated by the long and short dashed lines.

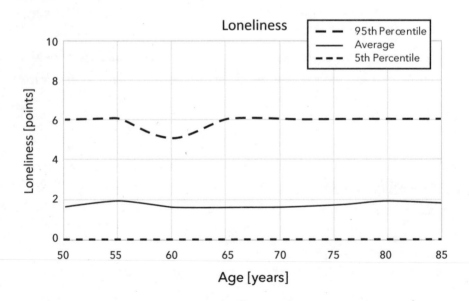

CENTER FOR EPIDEMIOLOGICAL STUDIES DEPRESSION SHORT FORM SCALE

This test gives a measure of depressive symptoms. Higher scores represent higher feelings of depression.

Put a circle around your response to each item then add the numbers together to yield your overall score for each section. Please do not leave any items blank.

	Rarely or none of the time (less than 1 day)	Some or a little of the time (1–2 days)	Occasionally or a moderate amount of time (3–4 days)	All of the time (5–7 days)
I felt depressed.	0	1	2	3
I felt everything I did was an effort.	0	1	2	3
My sleep was restless.	0	1	2	3
I was happy.	3	2	1	0
I felt lonely.	0	1	2	3

I enjoyed life.	3	2	1	0

I felt sad.	0	1	2	3

I could not 'get going'.	0	1	2	3

Total: _____

How do you compare?

Find your age along the horizontal axis and total score along the vertical axis to see where you are on the scale. Nearest to the solid black line is average; towards the long dashed line (95th percentile) is above and towards the short dashed line (5th percentile) below average. Ninety per cent of people are between the limits indicated by the short and long dashed lines.

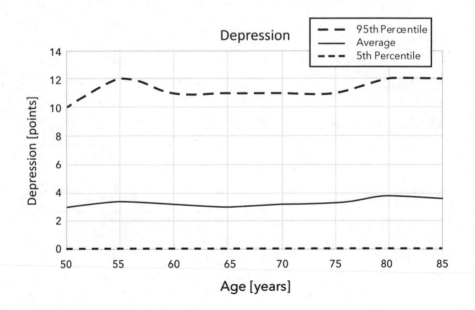

ONE LEG STAND

This test gives a measure of balance. Longer times represent better balance indicative of younger biological age. Ensure you perform this test on a stable surface.

One leg stand with eyes open
Stand on one leg and raise the other leg off the ground a few inches. Stand for as long as you can for up to 30 seconds. Your arms are free to move but make sure you do not hook the leg around your other leg, or rest it on it. You can choose either leg for this test.

One leg stand with eyes closed
Only perform this part of the test if you can complete the one leg stand with eyes open for five seconds or longer.

Close your eyes, lean your weight onto one leg and raise the other leg off the ground a few inches for as long as you can for up to 30 seconds. Your arms are free to move but make sure you do not hook the leg around your other leg, or rest it on it. You can choose either leg for this test and you do not have to use the same leg you used in the eyes open part of the test.

Record your time in seconds from the one leg stand with eyes closed.

Time (seconds): _____

How do you compare?

Find your age along the horizontal axis and total score in seconds along the vertical axis. The black line is the average.

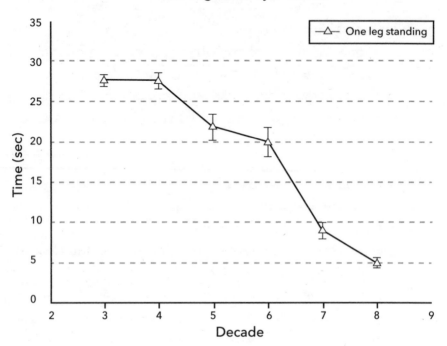

One Leg Stand eyes closed

Data from Luc Vereeck, Floris Wuyts, Steven Truijen and Paul Van de Heyning (2008) Clinical assessment of balance: Normative data, and gender and age effects, International Journal of Audiology, 47:2, 67-75, DOI: 10.1080/14992020701689688

References

CHAPTER 1

11   **We can measure epigenetics from blood samples and use the results to better understand why some people … live longer healthy lives:** Stringhini, S., et al., *Socioeconomic status, non-communicable disease risk factors, and walking speed in older adults: multi-cohort population based study.* BMJ, 2018. **360**: p. k1046.

McCrory, C., Kenny, R.A., et al., *The lasting legacy of childhood adversity for disease risk in later life.* Health Psychol, 2015. **34**(7): p. 687-96.

Stringhini, S., et al., *Socioeconomic status and the 25 × 25 risk factors as determinants of premature mortality: a multicohort study and meta-analysis of 1·7 million men and women.* The Lancet, 2017. **389**(10075): p. 1229-1237.

12   **Furthermore, we now understand a lot about the switching on and off of genes:** Chignon, A., et al., *Single-cell expression and Mendelian randomization analyses identify blood genes associated with lifespan and chronic diseases.* Commun Biol, 2020. **3**(1): p. 206.

12   **In species such as the worm, a small change in the DAF2 gene doubles lifespan:** Kenyon, C.J., *The genetics of ageing.* Nature, 2010. **464**(7288): p. 504-12.

13   **Diet, obesity and exercise and caloric restriction influence the DAF2 gene:** Milman, S., et al., *Low insulin-like growth factor-1 level predicts survival in humans with exceptional longevity.* Aging Cell, 2014. **13**(4): p. 769-771.

13   **… new 'clocks', which use different combinations of measures of methylation, continue to be discovered and tested for precision:** El Khoury, L.Y., et al.,

Systematic underestimation of the epigenetic clock and age acceleration in older subjects. Genome Biology, 2019. **20**(1): p. 283.

13   **No clock is as yet precise enough to clearly measure an individual's biological age:** McCrory, C., Kenny, R. A., et al., *Association of 4 epigenetic clocks with measures of functional health, cognition, and all-cause mortality in The Irish Longitudinal Study on Ageing (TILDA).* bioRxiv, 2020: p. 2020.04.27.063164.

Stringhini, S., et al., *Socioeconomic status, non-communicable disease risk factors, and walking speed in older adults*

McCrory, C., Kenny, R. A., et al., *The lasting legacy of childhood adversity for disease risk in later life*

Stringhini, S., et al., *Socioeconomic status and the 25 × 25 risk factors as determinants of premature mortality*

13   **... the epigenetic clock allows for calculation of the difference between chronological and biological age:** Belsky, D., et al., *Quantification of the pace of biological aging in humans through a blood test: a DNA methylation algorithm.* bioRxiv, 2020: p. 2020.02.05.927434.

13   **There has been recent hype surrounding this and there are now products on the market that claim to accurately determine biological age:** Mouratidis, Y. *We Are More Than Our DNA.* [Science 2018 Nov 17, 2018 July 16, 2020]; Available from: https://www.forbes.com/sites/yiannismouratidis/2018/11/17/we-are-more-than-our-dna/#385d42a52e9c.

14   **... they should be approached with caution ... as yet, the methods ... do not take into consideration all of the complex network of factors that influence the ageing process:** McCrory, C., Kenny, R. A., et al., *Epigenetic Clocks and Allostatic Load Reveal Potential Sex-Specific Drivers of Biological Aging.* J Gerontol A Biol Sci Med Sci, 2020. **75**(3): p. 495-503.

14   **Age acceleration:** Marioni, R.E., et al., *DNA methylation age of blood predicts all-cause mortality in later life.* Genome Biol, 2015; **16**(1): 25.

14   **Persistent stress and mood change ... state they create:** Lupien, S.J., et al., *Stress-induced declarative memory impairment in healthy elderly subjects: relationship to cortisol reactivity.* J Clin Endocrinol Metab, 1997. **82**(7): p. 2070-5.

Lupien, S.J., et al., *Effects of stress throughout the lifespan on the brain, behaviour and cognition.* Nat Rev Neurosci, 2009. **10**(6): p. 434-45.

14   **A well-known New Zealand study ... their ageing attitudes:** Caspi, A., et al., *Longitudinal Assessment of Mental Health Disorders and Comorbidities Across 4 Decades Among Participants in the Dunedin Birth Cohort Study.* JAMA Netw Open, 2020 Apr; **3**(4): p. e203221-e203221.

16   **For example, they had poorer balance, unable to stand on one leg for as long as slow agers; had clumsier fine-motor skills, when tested by placing small objects into holes on a pegboard, and had weaker grip strength:** Elliott, M.L., et al., *Brain-age in midlife is associated with accelerated biological aging and cognitive decline in a longitudinal birth cohort.* Mol Psychiatry, 2019 Dec 10:10.1038/s41380-019-0626-7.

Belsky, D.W., et al., *Eleven Telomere, Epigenetic Clock, and Biomarker-Composite Quantifications of Biological Aging: Do They Measure the Same Thing?* Am J Epidemiol, 2018. **187**(6): p. 1220-1230.

Elliott, M.L., et al., *Disparities in the pace of biological aging among midlife adults of the same chronological age have implications for future frailty risk and policy.* Nat Aging, 2021. **1**(3): p. 295-308.

Belsky, D., et al., *Quantification of the pace of biological aging in humans through a blood test*

Caspi, A., et al., *Longitudinal Assessment of Mental Health Disorders and Comorbidities*

16   **This shared starting point enables us to draw conclusions about brain vessels from eye vessels in adults:** Shalev, I., et al., *Retinal vessel caliber and lifelong neuropsychological functioning: retinal imaging as an investigative tool for cognitive epidemiology.* Psychol Sci, 2013. **24**(7): p. 1198-207.

19   **Changes detected in retinal photographs predict future stroke and vascular dementia:** Wong, T.Y. and P. Mitchell, *Hypertensive retinopathy.* N Engl J Med, 2004. **351**(22): p. 2310-7.

Ikram, M.A., et al., *The Rotterdam Study: 2018 update on objectives, design and main results.* Eur J Epidemiol, 2017. **32**(9): p. 807-850.

16   **Young adults in the Dunedin Study ... were at a higher risk of stroke and dementia later on in life:** Nolan, J.M., Kenny, R.A., et al., *Education is positively associated with macular pigment: the Irish Longitudinal Study on Ageing (TILDA).* Invest Ophthalmol Vis Sci, 2012. **53**(12): p. 7855-61.

Connolly, E., Kenny, R.A., et al., *Prevalence of age-related macular degeneration associated genetic risk factors and 4-year progression data in the Irish population.* Br J Ophthalmol, 2018. **102**(12): p. 1691-1695.

Feeney, J., Kenny, R.A., et al., *Low macular pigment optical density is associated with lower cognitive performance in a large, population-based sample of older adults.* Neurobiol Aging, 2013. **34**(11): p. 2449-56.

17   **The almost 12-year biological age difference was due, in the main, to adverse experiences in youth:** Belsky, D.W., *Reply to Newman: Quantification of biological aging in young adults is not the same thing as the onset of obesity.* Proc Natl Acad Sci USA, 2015. **112**(52): E7164-E7165.

17   … in the main, these resilient participants had positive perceptions, positive attitudes and optimism despite adverse circumstances: Snowdon, D., *Aging with Grace: What the Nun Study Teaches Us About Leading Longer, Healthier, and More Meaningful Lives*. 2002: Bantam.

17   But a number of studies from our group and others have now confirmed that we are 'as young as we feel', independent of disease status: Weiss, D. and F. Lang, *"They" Are Old But "I" Feel Younger: Age-Group Dissociation as a Self-Protective Strategy in Old Age*. Psychol Aging, 2012. **27**: p. 153-63.

17   … a positive attitude towards ageing changes … and epigenetics of the cell: Wurm, S. and Y. Benyamini, *Optimism buffers the detrimental effect of negative self-perceptions of ageing on physical and mental health*. Psychol Health, 2014. **29**(7): p. 832-48.

Wurm, S., et al., *How do negative self-perceptions of aging become a self-fulfilling prophecy?* Psychol Aging, 2013. **28**(4): p. 1088-97.

17   … people who feel close to or at their chronological age are more likely to develop physical frailty and poor brain health … than people who claim to feel younger than their chronological age: Robertson, D.A., Kenny, R.A., et al., *Negative perceptions of aging and decline in walking speed: a self-fulfilling prophecy*. PLoS One, 2015. **10**(4): e0123260.

Robertson, D.A. and R.A. Kenny, *Negative perceptions of aging modify the association between frailty and cognitive function in older adults*. Pers Individ Differ, 2016. **100**: 120-125.

Robertson, D.A., B.L. King-Kallimanis, and Kenny, R. A., *Negative perceptions of aging predict longitudinal decline in cognitive function*. Psychol Aging, 2016. **31**(1): p. 71-81.

McGarrigle C, Ward M, and Kenny, R.A., (In Press). *Negative Ageing Perceptions and Cognitive and Functional Decline: Are You As Old As You Feel?* JAGS.

18   When perceptions are negative, they result in declines in self-confidence, self-esteem … declines in physical and brain health.

Weiss, D. and F. Lang, *"They" Are Old But "I" Feel Younger: Age-Group Dissociation as a Self-Protective Strategy*

Wurm, S. and Y. Benyamini, *Optimism buffers the detrimental effect of negative self-perceptions of ageing on physical and mental health*

Wurm, S., et al., *How do negative self-perceptions of aging become a self-fulfilling prophecy?*

18   Negative perceptions make it more likely that someone will experience diseases such as heart disease, suffer heart attacks in later life and die early: Levy, B.R., et al., *Reducing cardiovascular stress with positive self-stereotypes of aging*. J Gerontol B Psychol Sci Soc Sci, 2000. **55**(4): p. P205-P213.

Levy, B.R., et al., *Age stereotypes held earlier in life predict cardiovascular events in later life.* Psychol Sci, 2009. **20**(3): p. 296-298.

Lang, P.O., J.P. Michel, and D. Zekry, *Frailty syndrome: a transitional state in a dynamic process.* Gerontology, 2009. **55**(5): p. 539-49.

18   … how quickly ageing perceptions can change an individual's physiology … included, 'Alzheimer's', 'confused', 'decline', 'decrepit', 'dementia', 'dependent', 'diseases', 'dying', 'forgets', 'incompetent', 'misplaces' and 'senile': Levy, B., *Improving memory in old age through implicit self-stereotyping.* J Pers Soc Psychol, 1996. **71**(6): p. 1092-1107.

Levy, B.R., et al., *Subliminal strengthening: improving older individuals' physical function over time with an implicit-age-stereotype intervention.* Psychol Sci, 2014. **25**(12): p. 2127-35.

18   **Those exposed to the negative stereotypes demonstrated unwanted excessive physiological responses, with higher blood … positive ageing stereotypes … led to a more modest physiological responses to stress.**

Levy, B.R., et al., *Reducing cardiovascular stress with positive self-stereotypes of aging.*

Levy, B., *Improving memory in old age through implicit self-stereotyping.*

19   **The more these older adults agreed with the negative statements like the first two … more likely they were to experience accelerated physical and cognitive ageing over the following eight years:** Robertson, D.A., Kenny, R.A., et al., *Negative perceptions of aging and decline in walking speed.*

Robertson, D.A. and Kenny R.A., *Negative perceptions of aging modify the association between frailty and cognitive function in older adults.*

Robertson, D.A., B.L. King-Kallimanis, and Kenny R.A., *Negative perceptions of aging predict longitudinal decline in cognitive function.*

19   **… showed that negative attitudes affected how different health conditions interact … frail participants with positive attitudes had the same level of mental ability as their non-frail peers:**   Robertson, D.A. and R.A. Kenny, *Negative perceptions of aging modify the association between frailty and cognitive function in older adults.*

20   **… underestimated how much sexual activity is linked to ageing perceptions … less likely to consider themselves old and less likely to believe that ageing has negative consequences:** *Sexual activity in the over 50s population in Ireland.* Orr, J., McGarrigle, C., Kenny, R.A., On behalf of the TILDA team February 2017 Copyright © The Irish Longitudinal Study on Ageing 2017 The Irish Longitudinal Study on Ageing Trinity College Dublin. https://tilda.tcd.ie/publications/reports/pdf/Report_SexualActivity.pdf.

Orr, J., R. Layte, N. O'Leary Kenny, R. A.,, *Sexual Activity and Relationship Quality in Middle and Older Age: Findings From The Irish Longitudinal Study on Ageing (TILDA)*. J Gerontol B Psychol Sci Soc Sci, 2019. **74**(2): p. 287-297.

20   **… older adults with negative attitudes about ageing live 7.5 years less … because of higher rates of heart diseases:** Levy, B., *Stereotype Embodiment:A Psychosocial Approach to Aging.* Curr Dir Psychol Sci, 2009 Dec 1; **18**(6): 332-336.

Jang, Y., L.W. Poon, and P. Martin, *Individual Differences in the Effects of Disease and Disability on Depressive Symptoms: The Role of Age and Subjective Health.* Int J Aging Hum Dev, 2004. **59**(2): p. 125-137.

Kim, S.H., *Older people's expectations regarding ageing, health-promoting behaviour and health status.* J Adv Nurs, 2009. **65**(1): p. 84-91.

Moor, C., et al., *Personality, aging self-perceptions, and subjective health: a mediation model.* Int J Aging Hum Dev, 2006. **63**(3): p. 241-57.

Levy, B.R., et al., *Reducing cardiovascular stress with positive self-stereotypes of aging.*

Levy, B.R., et al., *Age stereotypes held earlier in life predict cardiovascular events in later life.*

Levy, B., *Improving memory in old age through implicit self-stereotyping.*

20   **… we were in a position to show that perceptions independently affect early death:** Robertson, D.A., Kenny, R.A., et al., *Negative perceptions of aging and decline in walking speed.*

Robertson, D.A. and Kenny, R.A., *Negative perceptions of aging modify the association between frailty and cognitive function in older adults.*

Robertson, D.A., B.L. King-Kallimanis, and Kenny R.A., *Negative perceptions of aging predict longitudinal decline in cognitive function.*

McGarrigle C, Ward M, and Kenny, R.A., *Negative Ageing Perceptions and Cognitive and Functional Decline*

21   *As Young as You Feel* **is a 1951 comedy film … content with the outcome:** Wikipedia contributors. *As Young As You Feel.* [2020 10 May 2020 July 16, 2020]; Available from: https://en.wikipedia.org/w/index.php?title=As_Young_as_You_Feel&oldid=955839774

21   **… mandatory retirement allow employers … but rather a voluntary withdrawal from the work force at the age that best suits an individual's abilities, interests and career plans:** Till von Wachter, *The End of Mandatory Retirement in the US: Effects on Retirement and Implicit Contracts.* 2002: Columbia University. p. 60.

22   **Many European countries still have mandatory retirement for workers in the public sector:** Aegon Centre for Longevity and Retirement (ACLR) Survey. *Aegon*

Retirement Readiness Survey 2015: Inspiring a World of Habitual Savers. [2015 May 27, 2015 July 16, 2020]; Available from: https://www.aegon.com/research/reports/annual/aegon-retirement-readiness-survey-2015-inspiring-a-world-of-habitual-savers/

22   **Two thirds of EU citizens prefer to combine a part-time job and partial pension than to fully retire:** Eurofound, *European Quality of Life Survey 2016: Quality of Life, quality of public services, and quality of society,.* 2017: Publications Office of the European Union, Luxembourg. p. 122.

22   **Surveys from several European countries and the United States … less stress and greater life satisfaction, on average, than younger workers:** Nikolova, M. and C. Graham, *Employment, late-life work, retirement, and well-being in Europe and the United States.* IZA J Labor Stud 3, 5 (2014).

22   **Being able to choose when to stop working is important and affects both life satisfaction and ageing perceptions:** Walker, J.W. and H.L. Lazer, *The End of Mandatory Retirement: Implications for Management.* 1978, Chichester, New York: Wiley & Sons.

      OECD, *Pensions at a Glance 2017: OECD and G20 Indicators.* 2017, OECD Publishing, Paris.

22   **Unfortunately, mandatory retirement chimes with other negative societal attitudes … older adults as physically weak, forgetful, stubborn and selfish:** Lupien, S.J. and N. Wan, *Successful ageing: from cell to self.* Philos Trans R Soc London (Biol), 2004. **359**(1449): p. 1413-1426.

23   **… little objective medical or psychological evidence to support these commonly accepted 'truths':** World Health Organization. *Ageism.* [2020 July 16, 2020]; Available from: https://www.who.int/ageing/ageism/en/

23   **Only a small minority of older adults are physically, cognitively or mentally impaired … continues to get better after 50:** Layte, R., E. Sexton, G. Savva, Kenny, R. A., *Quality of life in older age: evidence from an Irish cohort study.* J Am Geriatr Soc, 2013. **61 Suppl 2**: p. S299-305.

23   **'He told me I was too old for that':** Royal Society for Public Health (RSPH), *That Age Old Question: How Attitudes To Ageing Affect Our Health and Wellbeing.* 2018: RSPH, London. .

      Abrams, D., Eilola T, and H. Swift, *Attitudes to age in Britain 2004-2008.* 2009, University of Kent: UK.

23   **A 2018 review … Only a third of Portuguese, Swiss and Germans said that they have older friends:** ESS9. *European Social Survey 2018.* [2018 July 30, 2020]; Available from: https://www.europeansocialsurvey.org/data/download.html?r=9

23   **... older adults are less likely to receive the same treatment simply because of age:** Jackson, S., R. Hackett, and A. Steptoe, *Associations between age discrimination and health and wellbeing: cross-sectional and prospective analysis of the English Longitudinal Study of Ageing.* Lancet Public Health, 2019:e200-e208.

24   **'overburdened service is no excuse for discrimination that would result in a cull of older people':** Hill, A. *Favouring young over old in COVID-19 treatment justifiable, says ethicist.* [2020 22 April, [2020 July 30, 2020]; Available from: https://www.theguardian.com/world/2020/apr/22/favouring-young-over-old-in-covid-19-treatment-justifiable-says-ethicist

25   **... almost 77 million babies were born in the United States ... boomers are increasingly likely to be healthy enough to run marathons, build houses and even start new businesses:** Chappelow, J. *Baby Boomer.* [Economics 2020 Feb 28, 2020 July 30, 2020]; Available from: https://www.investopedia.com/terms/b/baby_boomer.asp

26   **The Social Progress Index:** Porter M.E., Stern S, and Green M, *The Social Progress Index 2017.* 2017: Washington DC.

27   **Winters are long and dark in Denmark ... gathering of friends of all ages:** Parkinson, J., *A heart-warming lesson from Denmark.* 2015.

27   **Our language matters and ageism is exemplified by language and terminology:** Avers, D., et al., *Use of the Term "Elderly".* J Geriatr Phys Ther, 2011. **34**(4): p. 153-154.

27   **Certain words can be convenient but promote stereotyping through their generalisation ... the term 'elderly' for a person who is robust and independent as well as for a person who is frail and dependent says little about the individual:** Sarkisian, C.A., et al., *The relationship between expectations for aging and physical activity among older adults.* J Gen Intern Med, 2005. **20**(10): p. 911-5.

Sarkisian, C.A., et al., *Development, reliability, and validity of the expectations regarding aging (ERA-38) survey.* Gerontologist, 2002. **42**(4): p. 534-42.

Sarkisian, C.A., et al., *Correlates of attributing new disability to old age. Study of Osteoporotic Fractures Research Group.* J Am Geriatr Soc, 2001. **49**(2): p. 134-41.

Kim, S.H., *Older people's expectations regarding ageing, health-promoting behaviour and health status.*

27   **Consider how often you heard references to 'elderly' persons or 'the elderly' during the recent Covid crisis:** Palmore, E., *Ageism: Negative and Positive,.* 2nd ed. 1999: Springer Publishing Company.

28   **Ageist terms diminish older adults ... results in less care, less robust care and negatively affects outcomes:** Nemmers, T.M., *The Influence of Ageism and Ageist Stereotypes on the Elderly.* Phys Occup Ther Geriatr, 2005. **22**(4): p. 11-20.

28   **In a European survey, older individuals displayed a preference for 'older' or 'senior' and strongly rejected the terms 'aged', 'old' and – most strongly – 'elderly':** European Commission DG for Employment Social Affairs and Inclusion and DG Communication. *Special Eurobarometer 378 on Active ageing.* [2012 17 May 2012 September 9, 2020]; Available from: https://ec.europa.eu/eip/ageing/library/special-eurobarometer-378-active-ageing_en

Walker, A. and G.B.E. Gemeinschaften, *Age and attitudes: main results from a Eurobarometer survey.* 1993: Commission of the European Communities.

28   **... the United Nations Committee on Economic Social and Cultural Rights of Older Persons rejected the term 'elderly':** UN Committee on Economic Social and Cultural Rights (CESCR), *General Comment No. 6: The Economic, Social and Cultural Rights of Older Persons.* 1995. p. 11.

28   **... the International Longevity Centre recommended the term 'older adults' over 'senior' and 'elderly':** Dahmen, N. and R. Cozma, *Media takes: on aging.* 2008: International Longevity Center (USA) (ILC).

28   **... using terms that are precise, accurate, value-free and that older adults prefer makes good sense:** Kleinspehn-Ammerlahn, A., D. Kotter-Grühn, and J. Smith, *Self-perceptions of aging: do subjective age and satisfaction with aging change during old age?* J Gerontol B Psychol Sci Soc Sci, 2008. **63**(6): p. P377-85.

Kotter-Grühn, D., et al., *Self-perceptions of aging predict mortality and change with approaching death: 16-year longitudinal results from the Berlin Aging Study.* Psychol Aging, 2009. **24**(3): p. 654-67.

Levy, B.R. and L.M. Myers, *Preventive health behaviors influenced by self-perceptions of aging.* Prev Med, 2004. **39**(3): p. 625-9.

28   **... the 678 sisters of the Order of Notre Dame in Mankato, Minneapolis, USA, who agreed to take part in David Snowdon's longitudinal study in 1986 ... the influences of lifelong health and life experiences on the brain were mapped:** Tomasulo, D., *Learned Hopefulness: The Power of Positivity to Overcome Depression.* 2020: New Harbinger Publications. 192.

30   **Positive attitudes provide a type of inoculation to brain pathology:** Tomasulo, D. *Proof Positive: Can Heaven Help Us? The Nun Study - Afterlife.* 2010 13 May 2021]; Available from: https://psychcentral.com/blog/proof-positive-can-heaven-help-us-the-nun-study-afterlife#1

CHAPTER 2

33   **... the Blue Zones grew out of studies published in 2004:** Poulain, M., et al., *Identification of a geographic area characterized by extreme longevity in the Sardinia island: the AKEA study.* Exp Gerontol, 2004. **39**(9): p. 1423-9.

Poulain, M., A. Herm, and G. Pes, *The Blue Zones: areas of exceptional longevity around the world*. Vienna Yearbook of Population Research, 2013. **11**: p. 87-108.

34   ... scientists began to offer explanations for why these populations live healthier and longer lives: Buettner, D., *The Blue Zones. Lessons for living longer from the people who've lived the longest.* First Paperbacked. ed. 2009, Washington DC: National Geographic.

34   ... having a purpose in life makes us healthier, happier and, amazingly, can add up to seven years of extra life: Hill, P.L. and N.A. Turiano, *Purpose in Life as a Predictor of Mortality Across Adulthood.* Psychol Sci, 2014. **25**(7): p. 1482-1486.

35   In the case of the Adventists, their 'purpose' is being part of a faith-based community: Wallace, L.E., et al., *Does Religion Stave Off the Grave? Religious Affiliation in One's Obituary and Longevity.* Soc Psychol Personal Sci, 2019. **10**(5): p. 662-670.

36   *Overlap in healthy behaviours in three of the Blue Zones:* Buettner, D., *The Secrets of a Long Life*, in *National Geographic*. 2005, National Geographic.

         Wikipedia contributors. *Okinawa Island.* [2020 21 July 2020 July 28, 2020]; Available from: https://en.wikipedia.org/w/index.php?title=Okinawa Island&oldid=968792880

         Wikipedia contributors. *Icaria.* [2020 6 July 2020 July 28, 2020]; Available from: https://en.wikipedia.org/w/index.php?title=Icaria&oldid=966277626

37   ... the physician Alexander Leaf gave a detailed account of his journeys to communities of purported long-living people: Leaf, A., *Every day is a gift when you are over 100.*, in *National Geographic Magazine. Vol 143. No. 1, pp. 92-119.* 1973, National Geographic Society.: Washing D.C. p. 92-119.

37   ... unfortunately... Leaf acknowledged that there was no substantive objective evidence for longevity in the village of Vilcabamba: Leaf, A., *Statement Regarding the Purported Longevous Peoples of Vilcabamba*, in *In Controversial Issues in Gerontology, ed by H. Hershow.* 1981, Springer. p.25-26: New York. p. 25-26.

37   Further studies confirmed that none of the aforementioned areas survived scrutiny: Mazess, R.B. and S.H. Forman, *Longevity and age exaggeration in Vilcabamba, Ecuador.* J Gerontol, 1979. **34**(1): p. 94-8.

40   Calment's story was also challenged by them in a manuscript posted on ResearchGate.net: Zak, N, *Jeanne Calment: the secret of longevity.* 2018. DOI: 10.13140/RG.2.2.29345.04964.

40   ... they contested that it was mathematically impossible to live to Jeanne Calment advanced age: Zak, N., *Evidence That Jeanne Calment Died in 1934–Not 1997.* Rejuvenation Res, 2019. **22**(1): p. 3-12.

40 … **the claim was incorrect and was followed up a year later … discrediting Zak and Novoselov:** Robine, J.M., et al., *The Real Facts Supporting Jeanne Calment as the Oldest Ever Human.* J Gerontol A Biol Sci Med Sci, 2019. **74**(Supplement_1): p. S13-S20.

Robine, J.M., Allard M, *Validation of the exceptional longevity case of a 120 years old woman.*, in *Facts and Research in Gerontology. pp363-367.*

Desjardins, B., *Validation of extreme longevity cases in the Past: The French-Canadian Experience.*, in *Validation of Exceptional Longevity*, B. Jeune and J.W. Vaupel, Editors. 1999, Odense University Press: Denmark.

41 … **and significantly extend the life of a mouse:** Beyea, J.A., et al., *Growth hormone (GH) receptor knockout mice reveal actions of GH in lung development.* Proteomics, 2006. **6**(1): p. 341-348.

42 … **the theory that is most strongly embedded in popular belief:** Stibich, M. *What is the genetic theory of aging? How genes affect aging and how you may "alter" your genes.* [2020 January 26, 2020 April 1, 2020.]; Available from: https://www.verywellhealth.com/the-genetic-theory-of-aging-2224222

43 … **genes contribute to only 20–30 per cent of the variation in survival up to 80 years and genetic factors play a much stronger role in long living beyond 80:** Zeliadt N. *Live Long and Proper: Genetic Factors Associated with Increased Longevity Identified.* [2010 July 1, 2010 July 28, 2020]; Available from: https://www.scientificamerican.com/article/genetic-factors-associated-with-increased-longevity-identified/

44 … **several factors influenced appearance and facial ageing, including smoking and excessive sun exposure … in women over 40, a heavier body weight was associated with a more youthful look:** Parker-Pope, T. *Twins and the wrinkles of aging.* [2009 Feb 5 April 2, 2020.]; Available from: https://well.blogs.nytimes.com/2009/02/05/twin-studies-explain-wrinkles-of-aging/

45 … **lots of external factors other than genes contributed to an older appearance in identical twins:** Dorshkind, K., E. Montecino-Rodriguez, and R.A. Signer, *The ageing immune system: is it ever too old to become young again?* Nat Rev Immunol, 2009. **9**(1): p. 57-62.

Gudmundsson, H., et al., *Inheritance of human longevity in Iceland.* Eur J Hum, 2000. **8**(10): p. 743-749.

Sebastiani, P., et al., *Genetic signatures of exceptional longevity in humans.* PLoS One, 2012; **7**(1): e29848.

Puca, A.A., et al., *A genome-wide scan for linkage to human exceptional longevity identifies a locus on chromosome 4.* Proc Natl Acad Sci U S A, 2001. **98**(18): p. 10505-8.

47 **Proteins that function as recycling trucks … are also switched on and off by instructions from the nucleus:** Stibich, M. *What is the genetic theory of aging?*

21

48   **Ohsumi discovered how autophagy works ... Work is now focusing on ways to manipulate autophagy:** Kumsta, C., et al., *The autophagy receptor p62/SQST-1 promotes proteostasis and longevity in C. elegans by inducing autophagy.* Nat Commun, 2019. **10**(1): 5648.

48   **Another theory is that we are programmed to age:** Jin, K., *Modern Biological Theories of Aging.* Aging Dis, 2010. **1**(2): p. 72-74.

49   **... people with faster resting heart rates die earlier:** Fox, K., et al., *Resting Heart Rate in Cardiovascular Disease.* Journal of the American College of Cardiology, 2007. **50**(9): p. 823-830.

50   **... most human studies of antioxidant supplements have not as yet shown the same dramatic effects:** Eldridge, L., *Free Radicals: Definition, Causes, Antioxidants, and Cancer - What Exactly Are Free Radicals and Why Are they Important?* February 02, 2020 Accessed Oct 18 2021; Available from https://www.verywellhealth.com/information-about-free-radicals-2249103

50   **... the potential to delay or reverse the effects of ageing on the immune system would have significant benefits:** European Centre for Disease Prevention and Control (ECDC). *COVID-19 pandemic.* [2020 July 28, 2020]; Available from: https://www.ecdc.europa.eu/en/covid-19/latest-evidence/epidemiology

51   **Understanding the cell changes that underlie the decline in immune function has advanced:** Dorshkind, K., E. Montecino-Rodriguez, and R.A. Signer, *The ageing immune system.*

51   **Such compression of diseases and disability would create financial gains:** Science Advice for Policy by European Academies (SAPEA), *Transforming the Future of Ageing.* Michel, JP., Kuh, D., Kenny, R.A., et al., 2019: Berlin.

51   **France had almost 150 years to adapt ... However, Brazil, China and India have had only 20 years:** World Health Organization, *World Report on Ageing and Health.* 2015, WHO.

52   **Slowing down the process of ageing by just seven years could cut in half diseases:** Olshansky, S.J., L. Hayflick, and B.A. Carnes, *Position statement on human aging.* J Gerontol A Biol Sci Med Sci, 2002. **57**(8): p. B292-7.

CHAPTER 3

56   **Roseto, Pennsylvania, was its own tiny self-sufficient world:** Gladwell, M., *Outliers: The Story of Success.* 2008: Penguin.

56   **Wolf thought that there might be something specific about the lifestyle of the town's residents:** Oransky, I., *Stewart Wolf.* The Lancet, 2005. **366**(9499): p. 1768.

58   **Wolf and Bruhn ... emphasised how residents had avoided the internalisation of stress by sharing resources, worries and emotions:** Wolf, S. and J.G. Bruhn, *The Power of Clan: Influence of Human Relationships on Heart Disease.* 1998: Routledge.

Grossman, R. and C. Leroux. *A New "Roseto Effect".* [1996 October 11, 1996 August 17, 2020]; Available from: https://www.chicagotribune.com/news/ct-xpm-1996-10-11-9610110254-story.html.

58   **... research in monkeys can often be translated into human observations and studies ... in what is known as a randomised control trial:** Mattison, J.A., et al., *Caloric restriction improves health and survival of rhesus monkeys.* Nat Commun 2017. **8**(1): p. 14063.

59   **... longevity among the macaques is linked to having strong social connections, including spending time together and time grooming each other's fur:** Christakis, N.A. and P.D. Allison, *Mortality after the hospitalization of a spouse.* N Engl J Med, 2006. **354**(7): p. 719-30.

Holt-Lunstad, J., T.B. Smith, and J.B. Layton, *Social relationships and mortality risk: a meta-analytic review.* PLoS Med, 2010 Jul 27; **7**(7): e1000316.

House, J.S., K.R. Landis, and D. Umberson, *Social relationships and health.* Science, 1988. **241**(4865): p. 540-5.

Seeman, T.E., *Social ties and health: the benefits of social integration.* Ann Epidemiol, 1996. **6**(5): p. 442-51.

59   **In adult female macaques, close female relatives are a proxy for friendships ... require less social support for 'protection':** Brent, L.J.N., A. Ruiz-Lambides, and M.L. Platt, *Family network size and survival across the lifespan of female macaques.* Proc Biol Sci, 2017. **284**(1854).

Ellis, S., et al., *Deconstructing sociality: the types of social connections that predict longevity in a group-living primate.* Proc Royal Soc B, 2019. **286**(1917): 20191991.

House, J.S., K.R. Landis, and D. Umberson, *Social relationships and health.*

60   **Social relationships are not just important in macaques but rather are associated with longer lifespan in many other affable species:** Archie, E.A., et al., *Social affiliation matters: both same-sex and opposite-sex relationships predict survival in wild female baboons.* Proc Royal Soc B, 2014. **281**(1793): 20141261.

Silk, J.B., et al., *Strong and consistent social bonds enhance the longevity of female baboons.* Curr Biol, 2010. **20**(15): p. 1359-61.

Stanton, M.A. and J. Mann, *Early social networks predict survival in wild bottlenose dolphins.* PLoS One, 2012; **7**(10): e47508.

Yee, J.R., et al., *Reciprocal affiliation among adolescent rats during a mild group stressor predicts mammary tumors and lifespan.* Psychosomatic medicine, 2008. **70**(9): p. 1050-1059.

60   **implying a common evolutionary basis across species for friendship** Brent, L.J., et al., *The neuroethology of friendship.* Ann N Y Acad Sci, 2014. **1316**(1): p. 1-17.

Almeling, L., et al., *Motivational Shifts in Aging Monkeys and the Origins of Social Selectivity.* Curr Biol, 2016. **26**(13): p. 1744-1749.

Brent, L.J.N., et al., *Ecological knowledge, leadership, and the evolution of menopause in killer whales.* Curr Biol, 2015. **25**(6): p. 746-750.

Nussey, D.H., et al., *Senescence in natural populations of animals: widespread evidence and its implications for bio-gerontology.* Ageing Res Rev, 2013. **12**(1): p. 214-25.

Holt-Lunstad, J., T.B. Smith, and J.B. Layton, *Social relationships and mortality risk.*

60   **... the overwhelming majority of studies to date have focused on links between sociality and longevity in older people:** Giles, L.C., et al., *Effect of social networks on 10 year survival in very old Australians: the Australian longitudinal study of aging.* J Epidemiol Community Health, 2005. **59**(7): p. 574-9.

Steptoe, A., et al., *Social isolation, loneliness, and all-cause mortality in older men and women.* Proc Natl Acad Sci U S A, 2013. **110**(15): p. 5797-801.

Luo, Y., et al., *Loneliness, health, and mortality in old age: a national longitudinal study.* Soc Sci Med, 2012. **74**(6): p. 907-14.

60   **In humans ... social network size is important for physical health:** Yang, Y.C., et al., *Social relationships and physiological determinants of longevity across the human life span.* Proc Natl Acad Sci USA, 2016. **113**(3): p. 578-583.

60   **... detail why social interactions matter and what types of social networks affect our health:** Berkman, L.F. and S.L. Syme, *Social networks, host resistance, and mortality: a nine-year follow-up study of Alameda County residents.* Am J Epidemiol, 1979. **109**(2): p. 186-204.

61   **... many longitudinal studies have reinforced the impact of social ties on death rates:** Christakis, N.A. and P.D. Allison, *Mortality after the hospitalization of a spouse.*

Holt-Lunstad, J., T.B. Smith, and J.B. Layton, *Social relationships and mortality risk.*

House, J.S., K.R. Landis, and D. Umberson, *Social relationships and health.*

Seeman, T.E., *Social ties and health.*

Giles, L.C., et al., *Effect of social networks on 10 year survival in very old Australians.*

Steptoe, A., et al., *Social isolation, loneliness, and all-cause mortality in older men and women.*

Luo, Y., et al., *Loneliness, health, and mortality in old age*

61   **The strength of the association between fibrinogen and social isolation was remarkable ... the same as for smoking:** Kim, D.A., et al., *Social connectedness is associated with fibrinogen level in a human social network.* Proc Biol Sci, 2016. **283**(1837): 20160958.

62   **These primate observations align with human observations in Roseto and other social network research:** Vandeleest, J.J., et al., *Social stability influences the association between adrenal responsiveness and hair cortisol concentrations in rhesus macaques.* Psychoneuroendocrinology, 2019. **100**: p. 164-171.

Capitanio, J.P., S. Cacioppo, and S.W. Cole, *Loneliness in monkeys: Neuroimmune mechanisms.* Curr Opin Behav Sci, 2019. **28**: p. 51-57.

62   **... described the human-like social gestures she observed ... [Sylvia] began offering to groom the peers she'd once scorned:** Denworth, L., *Friendship: The Evolution, Biology, and Extraordinary Power of Life's Fundamental Bond.* 2020: W. W. Norton & Company.

62   **... illustrates how friendship is hard-wired; it is not a choice or a luxury but rather a necessity:** Brent, L.J., et al., *Genetic origins of social networks in rhesus macaques.* Sci Rep, 2013. **3**: 1042.

Brent, L.J.N., J. Lehmann, and G. Ramos-Fernández, *Social network analysis in the study of nonhuman primates: a historical perspective.* American journal of primatology, 2011. **73**(8): p. 720-730.

62   **... puncture the stereotype that female friendships thrive on endless chats and male ones on side-by-side activity ... men reported being more satisfied:** Fehr, B., *Friendship Processes.* 1996: SAGE Publications, Inc: 1 edition.

63   **... many male friendships also require depth:** Denworth, L., *Friendship.*

63   **... somehow, among a myriad of possibilities, we manage to select as friends the people who resemble our kin:** Settle, J.E., et al., *Friendships Moderate an Association Between a Dopamine Gene Variant and Political Ideology.* J Politics, 2010. **72**(4): p. 1189-1198.

63   **... researchers who studied 5,000 pairs of adolescent friends ran a number of genetic comparisons:** Christakis, N.A. and J.H. Fowler, *Friendship and natural selection.* Proc Natl Acad Sci USA, 2014. **111**(Supplement 3): p. 10796-10801.

63   **... we are also more similar genetically to our spouses:** Domingue, B.W., et al., *Genetic and educational assortative mating among US adults.* Proc Natl Acad Sci USA, 2014. **111**(22): p. 7996-8000.

Christakis, N.A. and J.H. Fowler, *Friendship and natural selection.*

63   **Genes drive both friend choice and loneliness**: Fowler, J.H., J.E. Settle, and N.A. Christakis, *Correlated genotypes in friendship networks.* Proc Natl Acad Sci USA, 2011;108(5): p.1993-1997.

Cacioppo, J.T., J.H. Fowler, and N.A. Christakis, *Alone in the crowd: the structure and spread of loneliness in a large social network.* J Pers Soc Psychol, 2009. **97**(6): p. 977-991.

Christakis, N.A. and J.H. Fowler, *Friendship and natural selection.*

64   **There are a number of key strategies that will help with loneliness:** Murthy, V., *Together - The Healing Power of Human Connection in a Sometimes Lonely World.* 2020: Harper Wave.

64   **... more than 9 million people in Britain often or always feel lonely ... Loneliness is estimated to cost UK employers up to £3.5 billion annually:** Tara John. *How the World's First Loneliness Minister Will Tackle "the Sad Reality of Modern Life".* [2018 April 25, 2018 August 17, 2020]; Available from: https://time.com/5248016/tracey-crouch-uk-loneliness-minister/

65   **... a quarter of Irish adults feel lonely some of the time ... lonely people are also more likely to suffer from depression:** Ward M, Kenny, R.A., et al., *Loneliness and social isolation in the COVID-19 Pandemic among the over 70s: Data from The Irish Longitudinal Study on Ageing (TILDA) and ALONE.* 2020, TILDA, Trinity College Dublin.

65   **... in Japan was in 2000, when the corpse of a 69-year-old man was discovered:** Onishi, N. *A Generation in Japan Faces a Lonely Death.* [2017 Nov 30, 2017 August 17, 2020]; Available from: https://www.nytimes.com/2017/11/30/world/asia/japan-lonely-deaths-the-end.html

65   **In 2008, there were more than 2,200 reported lonely deaths in Tokyo ... reported that 20 per cent of the their jobs involved removing the belongings of people who had died lonely deaths:** Suzuki Hikaru, *Death and Dying in Contemporary Japan.* 1 ed. 2012: Routledge, 1 edition.

65   *Kodokushi* **mostly affects men who are 50 or older:** Wikipedia contributors. *Kodokushi.* [2020 4 August 2020 August 18, 2020]; Available from: https://en.wikipedia.org/w/index.php?title=Kodokushi&oldid=971219759

65   **Social isolation is increasing as older Japanese people are living alone more and more ... more likely to die alone and remain undiscovered:** Leng Leng Thang, *Generations in Touch: Linking the Old and Young ina Tokyo Neighborhood.* The Anthropology of Contemporary Issues. 2001: Cornell University Press.

66   **Victims of *kodokushi* have been described as 'slipping through the cracks':** Wikipedia contributors. *Kodokushi.*

66   **In a recent US survey of over 20,000 people aged 18 upwards:** Bruce, L.D., et al., *Loneliness in the United States: A 2018 National Panel Survey of Demographic,*

Structural, Cognitive, and Behavioral Characteristics. Am J Health Promot, 2019. **33**(8): p. 1123-1133.

66 **Household size is shrinking and there are now more single-person households in Europe than any other:** Eurostat. [2019 August, 19 2020]; Available from: https://ec.europa.eu/eurostat/statistics-explained/index.php?title=Household_composition_statistics

66 **How much effort we invest into relationships affects the level of support we receive from them:** Roberts, B.W., D. Wood, and J.L. Smith, *Evaluating Five Factor Theory and social investment perspectives on personality trait development.* J Res Pers, 2005. **39**(1): p. 166-184.

Carstensen, L.L., D.M. Isaacowitz, and S.T. Charles, *Taking time seriously: A theory of socioemotional selectivity.* Am Psychol, 1999. **54**(3): p. 165-181.

66 **Harmonious family relationships have a long history of endowing positive effects on people**: Solomon, B.C. and J.J. Jackson, *The Long Reach of One's Spouse:Spouses' Personality Influences Occupational Success.* Psychol Sci, 2014. **25**(12): p. 2189-2198.

Umberson, D., *Relationships between adult children and their parents: Psychological consequences for both generations.* J Marriage Fam, 1992. **54**(3): p. 664-674.

66 **... two large-scale analyses to understand the relative contributions of friends and family to good health and happiness:** Chopik, W.J., *Associations among relational values, support, health, and well-being across the adult lifespan.* Pers Relatsh, 2017. **24**(2): p. 408-422.

67 **The findings align well with other research on the overall and lasting benefits of close relationships:** House, J.S., K.R. Landis, and D. Umberson, *Social relationships and health.*

68 **... when friends and family are the source of strain, people experience more chronic illnesses:** Bearman, P.S. and J. Moody, *Suicide and friendships among American adolescents.* Am J Public Health, 2004. **94**(1): p. 89-95.

Christakis, N.A. and J.H. Fowler, *The spread of obesity in a large social network over 32 years.* N Engl J Med, 2007. **357**(4): p. 370-9.

Giles, L.C., et al., *Effect of social networks on 10 year survival in very old Australians*

68 **... social networks tend to decrease in size as we mature, we shift more attention and resources toward maintaining existing relationships:** Carstensen, L.L., D.M. Isaacowitz, and S.T. Charles, *Taking time seriously*

68 **... our interactions with friends stem from choice:** Giles, L.C., et al., *Effect of social networks on 10 year survival in very old Australians*

68   **On days when we positively interact with friends, we report greater happiness**: Sandstrom, G.M. and E.W. Dunn, *Social Interactions and Well-Being: The Surprising Power of Weak Ties*. Pers Soc Psychol Bull, 2014. **40**(7): p. 910-922.

Huxhold, O., M. Miche, and B. Schüz, *Benefits of having friends in older ages: differential effects of informal social activities on well-being in middle-aged and older adults*. J Gerontol B Psychol Sci Soc Sci, 2014. **69**(3): p. 366-75.

68   **Friendships are more closely tied to well-being**: Larson, R., R. Mannell, and J. Zuzanek, *Daily well-being of older adults with friends and family*. Psychology and Aging, 1986. **1**(2): p. 117-126.

68   **We should also bear this science in mind if dealing with future pandemics**: N. Clarke, R.A. Kenny, et al., *Altered lives in a time of crisis: The impact of the COVID-19 pandemic on the lives of older adults in Ireland Findings from The Irish Longitudinal Study on Ageing*. Dublin, 2021.

69   **Historically, large studies show that, on average, married people report greater happiness**: Lee, K.S. and H. Ono, *Marriage, Cohabitation, and Happiness: A Cross-National Analysis of 27 Countries*. J Marriage Fam, 2012. **74**(5): p. 953-972.

69   **The positive effects of marriage on happiness are reported by both women and men**: Diener, E., et al., *Similarity of the Relations between Marital Status and Subjective Well-Being Across Cultures*. J Cross Cult Psychol, 2000. **31**(4): p. 419-436.

69   **Happy people who get married still end up happier**: Stutzer, A. and B.S. Frey, *Does marriage make people happy, or do happy people get married?* J Socio Econ, 2006. **35**(2): p. 326-347.

69   **... marital satisfaction is a much stronger predictor of happiness than just being married**: Carr, D., et al., *Happy Marriage, Happy Life? Marital Quality and Subjective Well-being in Later Life*. J Marriage Fam, 2014. **76**(5): p. 930-948.

69   **Single people who elect to never marry but have strong social support through other means are certainly happy**: Hostetler, A.J., *Singlehood and Subjective Well-Being among Mature Gay Men: The Impact of Family, Friends, and of Being "Single by Choice"*. J GLBT Fam, 2012. **8**(4): p. 361-384.

69   **... being in a long-term, committed relationship ... is definitely good**: Bourassa, K.J., D.A. Sbarra, and M.A. Whisman, *Women in very low quality marriages gain life satisfaction following divorce*. J Fam Psychol, 2015. **29**(3): p. 490-499.

Dolan, P., *Happy Ever After: Escaping The Myth of The Perfect Life*. 2019: Allen Lane. 256.

70   **... people who enjoyed strong social bonds into their eighties were less likely to succumb to late-life cognitive decline and dementia**: Butler, R.N., F. Forette, and B.S. Greengross, *Maintaining cognitive health in an ageing society*. J R Soc Promot Health, 2004. **124**(3): p. 119-121.

70   **Researchers at Michigan State University tested which aspects of social relations were most associated with memory**: Zahodne, L.B., et al., *Social relations and age-related change in memory.* Psychol Aging, 2019. **34**(6): p. 751-765.

70   **Social engagement, connections with family and friends and participation in activities and organisations protects against poor cognitive functioning**: Fratiglioni, L., S. Paillard-Borg, and B. Winblad, *An active and socially integrated lifestyle in late life might protect against dementia.* Lancet Neurol, 2004. **3**(6): p. 343-53.

70   **... colleagues at University College, London conducted a large review of the published literature ... All [lifestyle components] seemed to share common pathways, rather than having specific separate pathways**: Hackett, R.A., et al., *Social engagement before and after dementia diagnosis in the English Longitudinal Study of Ageing.* PLoS One, 2019. **14**(8): p. e0220195.

71   **... an impoverished environment of loneliness and low activity in rats is linked to impaired brain function. The good news is that this is partly reversible**: Winocur, G., *Environmental influences on cognitive decline in aged rats.* Neurobiol Aging, 1998. **19**(6): p. 589-97.

Pham, T.M., et al., *Effects of environmental enrichment on cognitive function and hippocampal NGF in the non-handled rats.* Behav Brain Res, 1999. **103**(1): p. 63-70.

Pham, T.M., et al., *Environmental influences on brain neurotrophins in rats.* Pharmacol Biochem Behav, 2002. **73**(1): p. 167-175.

72   **... new brain cell formation and cognitive reserve pretty much covers most of our important brain functions**: Churchill, J.D., et al., *Exercise, experience and the aging brain.* Neurobiol Aging, 2002. **23**(5): p. 941-55.

72   **MRI brain scans confirm that people with higher cognitive reserve**: Scarmeas, N. and Y. Stern, *Cognitive reserve and lifestyle.* J Clin Exp Neuropsychol, 2003. **25**(5): p. 625-33.

72   **Social, mental and physical stimulation through friendships and relationships also act via the vascular system**: Skoog, I., et al., *15-year longitudinal study of blood pressure and dementia.* Lancet, 1996. **347**(9009): p. 1141-5.

de la Torre, J.C., *Alzheimer disease as a vascular disorder: nosological evidence.* Stroke, 2002. **33**(4): p. 1152-62.

Launer, L.J., *Demonstrating the case that AD is a vascular disease: epidemiologic evidence.* Ageing Res Rev, 2002. **1**(1): p. 61-77.

Fratiglioni, L., S. Paillard-Borg, and B. Winblad, *An active and socially integrated lifestyle in late life might protect against dementia.*

72   **A higher susceptibility to stress doubles the risk of dementia:** Yaffe, K., et al., *Posttraumatic stress disorder and risk of dementia among US veterans.* Arch Gen Psychiatry, 2010. **67**(6): p. 608-13.

CHAPTER 4

73   **We are wired to be happy … Laughter is a social behaviour:** Wellenzohn, S., R.T. Proyer, and W. Ruch, *Who Benefits From Humor-Based Positive Psychology Interventions? The Moderating Effects of Personality Traits and Sense of Humor.* Front Psychol, 2018. **9**: p. 821.

    O'Nions, E., et al., *Reduced Laughter Contagion in Boys at Risk for Psychopathy.* Curr Biol, 2017. **27**(19): p. 3049-3055 e4.

73   **You can actually tell the strength of a relationship between people from the tone and type of laughter:** Lavan, N., et al., *Flexible voices: Identity perception from variable vocal signals.* Psychon Bull Rev, 2019. **26**(1): p. 90-102.

    Lavan, N., S. Scott, and C. McGettigan, *Laugh Like You Mean It: Authenticity Modulates Acoustic, Physiological and Perceptual Properties of Laughter.* J Nonverbal Behav, 2016. **40**: p. 133-149

73   **… are all different and convey the type of relationship being shared:** Lavan, N., et al., *Neural correlates of the affective properties of spontaneous and volitional laughter types.* Neuropsychologia, 2017. **95**: p. 30-39.

74   **As well as being a feel-good action, laughter contributes to better health:** Goldstein, J.H., *A Laugh A Day.* The Sciences, 1982. **22**(6): p. 21-25.

74   **We are also more likely to laugh when there are other people around**: Cai, Q.C., et al., *Modulation of humor ratings of bad jokes by other people's laughter.* Current Biology, 2019. **29** (14): p. R677-R678.

74   **Friends spend on average 10 per cent of a conversation laughing:** Scott, S. *What do we know about laughter?* Huxley Summit 2017 Dec 2017; Available from: https://www.youtube.com/watch?v=Ow824i0nvRc.

74   **Laughter is therefore key to important social interactions … play meaningful physiological and psychological roles**: Scott, S. *Why we laugh [video file].* TED2015 2015 March Available from: https://www.ted.com/talks/sophie_scott_why_we_laugh?referrer=playlist-10_days_of_positive_thinking

    Scott, S. *What do we know about laughter?*

74   **By elevating mood, laughter reduces stress levels in all parties:** Savage, B.M., et al., *Humor, laughter, learning, and health! A brief review.* Adv Physiol Educ, 2017. **41**(3): p. 341-347.

74   **Dogs laugh … Even rats laugh:** Scott, S. *Why we laugh [video file].*

75   **The benefits of humour and laughter are well detailed through history, as early as the reign of Solomon:** Proverbs 17:22 NIV, *A cheerful heart is good medicine, but a crushed spirit dries up the bones,* in *the Bible.*

75   **Ancient Greek physicians, as an adjunct to therapy, prescribed a visit to the hall of comedians:** Kleisiaris, C.F., C. Sfakianakis, and I.V. Papathanasiou, *Health care practices in ancient Greece: The Hippocratic ideal.* J Med Ethics Hist Med, 2014. **7**: p. 6.

      Savage, B.M., et al., *Humor, laughter, learning, and health!*

75   **Early Native Americans used the impact of humour and laughter in healing:** Emmons, S.L., *A disarming laughter: The role of humor in tribal cultrues. An examination of humor in contemporary Native American literature and art.*, in *Department of English.* 2000, University of Oklahoma. p. 262.

75   **'Let the surgeon take care to regulate the whole regimen of the patient's life ...':** Clarke, C.C., *Henri De Mondeville.* Yale J Biol Med, 1931. **3**(6): p. 458-81.

75   **The English parson and scholar Robert Burton extended this practice by using humour:** Burton, R., *The Anatomy of Melancholy.* 1977, New York, United States: Vintage Books.

75   **Luther advised individuals with depression not to isolate themselves:** Wells, K., *Humor Therapy*, in *The Gale Encyclopedia of Alternative Medicine*, L. J, Editor. 2001, Thomson Gale: Detroit, MI. p. 1009-1010.

76   **When we laugh, we are using the intercostals:** Scott, S. *Why we laugh [video file].*

76   **Laughter provides a physical release ... It even provides a good workout for the immune system and the heart:** Scott, E. *How to Deal With Negative Emotions and Stress.* [Emotions 2020 April 30, 2020 June, 23 2020]; Available from: https://www.verywellmind.com/how-should-i-deal-with-negative-emotions-3144603

      Ghiadoni, L., et al., *Mental stress induces transient endothelial dysfunction in humans.* Circulation, 2000. **102**(20): p. 2473-8.

      Hayashi, T., et al., *Laughter up-regulates the genes related to NK cell activity in diabetes.* Biomed Res J, 2007. **28**(6): p. 281-285.

77   **Low cortisol stabilises blood sugars and insulin:** Savage, B.M., et al., *Humor, laughter, learning, and health!*

77   **A single one-hour episode of daily mirthful laughter lowered the rate of recurrent heart attack:** Berk, L., Tan, LG, Tan SA, *Mirthful Laughter, as Adjunct Therapy in Diabetic Care, Attenuates Catecholamines, Inflammatory Cytokines, C – reactive protein, and Myocardial Infarction Occurrence*, in *FASEB 2008.* 2008, Experimental Biology 2017 Meeting Abstracts: San Diego, CA.

77   **Laughter also increases endorphins:** Tan, S.A., et al., *Humor, as an adjunct therapy in cardiac rehabilitation, attenuates catecholamines and myocardial infarction recurrence.* Adv Mind Body Med, 2007. **22**(3-4): p. 8-12.

      Lavan, N., S. Scott, and C. McGettigan, *Laugh Like You Mean It.*

Cai, Q.C., et al., *Modulation of humor ratings of bad jokes by other people's laughter.*

77 **How much better to stimulate these systems through laughter:** Takahashi, K., et al., *The elevation of natural killer cell activity induced by laughter in a crossover designed study.* Int J Mol Med, 2001. **8**(6): p. 645-650.

77 **Endorphins are not just about pain and stress:** Scott, S. *Voluntary and Involuntary Mechanisms in Laughter Production and Perception.* in *Proceedings of Laughter Workshop* 2018. Sorbonne University: academia.eu.

Takahashi, K., et al., *The elevation of natural killer cell activity induced by laughter.*

77 **... boosting endorphins is particularly beneficial in older persons. High-stress hormones weaken our immune system:** Dillon, K.M., B. Minchoff, and K.H. Baker, *Positive emotional states and enhancement of the immune system.* Int J Psychiatry Med, 1985. **15**(1): p. 13-8.

Savage, B.M., et al., *Humor, laughter, learning, and health!*

Scott, E. *How to Deal With Negative Emotions and Stress.*

78 **Levels of good chemicals such as endorphins rose to as high as 87 per cent from the baseline level:** Berk, L.S., S.A. Tan, and D. Berk, *Cortisol and Catecholamine stress hormone decrease is associated with the behavior of perceptual anticipation of mirthful laughter.* The FASEB Journal, 2008. **22**(S1): p. 946.11-946.11.

78 **... laughter therapy works for patients with depression:** Bressington, D., et al., *The effects of group–based Laughter Yoga interventions on mental health in adults: A systematic review.* J Psychiatr Ment Health Nurs, 2018. **25**(8): p. 517-527.

78 **There are a number of websites where you can find out more about laughter therapy and laughter yoga:** Yim, J., *Therapeutic Benefits of Laughter in Mental Health: A Theoretical Review.* Tohoku J Exp Med, 2016. **239**(3): p. 243-9.

78 **... the number of times we laugh declines as we get older but the potential for physical and psychological advantage is still there:** Yoshikawa, Y., et al., *Beneficial effect of laughter therapy on physiological and psychological function in elders.* Nurs Open, 2019. **6**(1): p. 93-99.

78 **Purpose is a key psychological strength:** Ryff, C.D., *The Benefits of Purposeful Life Engagement on Later-Life Physical Function.* JAMA Psychiatry, 2017. **74**(10): p. 1046-1047.

78 **One of the first physicians to detail the value of purpose was a psychiatrist who spent three years as a prisoner in Nazi concentration camps ... the book has sold over 16 million copies:** Frankl, V.E., *Man's Search for Meaning.* 1959, Boston, MA, United States: Beacon Press.

80 **Purpose is about reflective activities:** Ryff, C.D., *The Benefits of Purposeful Life Engagement on Later-Life Physical Function.*

81 **An abundance of data shows that people who engage in volunteering are less depressed**: Ward, M., et al., The Irish Longitudinal Study on Ageing (TILDA), *TILDA Wave 4 Report: Wellbeing and Health in Ireland's over 50s 2009-2016*. 2018, Trinity College Dublin.

81 **Volunteers are needed in so many different domains:** Ward, M., S. Gibney, and I. Mosca, *Volunteering and social participation*, in *TILDA Wave 4 Report: Welbeing and health in Ireland's over 50s 2009-2016*. Kenny, R. A., 2018: Trinity College Dublin.

81 **Grandparenting provides purpose in a multiplex of ways**: Aassve, A., B. Arpino, and A. Goisis, *Grandparenting and mothers' labour force participation: A comparative analysis using the Generations and Gender Survey*. Demogr Res, 2012. **S11**(3): p. 53-84.

81 **Characteristic of many centenarians is a persistent sense of purpose:** Antonini, F.M., et al., *Physical performance and creative activities of centenarians*. Archives of Gerontology and Geriatrics, 2008. **46**(2): p. 253-261.

Katz, J., et al., *A Better Life: what older people with high support needs value*, I. Blood, Editor. 2011: Joseph Rowntree Foundation https://www.jrf.org.uk/report/better-life-what-older-people-high-support-needs-value

81 **Activities such as choir membership, gardening or a new academic degree, course or diploma ... provide a sense of purpose and positive psychological health benefits:** Cohen, G.D., et al., *The impact of professionally conducted cultural programs on the physical health, mental health, and social functioning of older adults*. Gerontologist, 2006. **46**(6): p. 726-34.

Nimrod, G., *Retirees' Leisure: Activities, Benefits, and their Contribution to Life Satisfaction*. Leisure Studies, 2007. **26**: 1, p. 65-80.

81 **Purpose is also amplified through creativity:** Price, K.A. and A.M. Tinker, *Creativity in later life*. Maturitas, 2014. **78**(4): p. 281-286.

81 **Neurological research shows that making art improves not just mood but also cognitive function:** Mclean, J., et al., *An Evidence Review of the impact of Participatory Arts on Older People*. 2011, Mental Health Foundation, London.

81 **... while brains inevitably age, creative abilities do not deteriorate:** Miller, B.L. and C.E. Hou, *Portraits of artists: emergence of visual creativity in dementia*. Arch Neurol, 2004. **61**(6): p. 842-4.

81 **Imagination and creativity flourish in later life:** Haier, R.J. and R.E. Jung, *Brain Imaging Studies of Intelligence and Creativity: What is the Picture for Education?* Roeper Review, 2008. **30**(3): p. 171-180.

82 **People involved in weekly art participation have better physical health:** Cohen, G.D., et al., *The impact of professionally conducted cultural programs on the physical health, mental health, and social functioning.*

Price, K.A. and A.M. Tinker, *Creativity in later life.*

83   **Our research clearly shows a positive relationship between religious practice, heart disease and death:** Orr, J., Kenny, R.A., et al., *Religious Attendance, Religious Importance, and the Pathways to Depressive Symptoms in Men and Women Aged 50 and Over Living in Ireland.* Res Aging, 2019. **41**(9): p. 891-911.

Central Statistics Office, *Census 2016 Results Profile 8 - Irish Travellers, Ethnicity and Religion* in *Census 2016 Results* C.S. Office, Editor. 2017: Dublin, Ireland.

Inglis, T., *Moral monopoly: The rise and fall of the Catholic Church in modern Ireland.* 1998: Univ College Dublin Press.

Chida, Y., A. Steptoe, and L.H. Powell, *Religiosity/spirituality and mortality. A systematic quantitative review.* Psychother Psychosom, 2009. **78**(2): p. 81-90.

83   **... with lower blood pressure and better immunity in religious Irish adults:** Orr, J., Kenny, R.A., et al., *Religious Attendance, Religious Importance, and the Pathways to Depressive Symptoms in Men and Women Aged 50 and Over Living in Ireland.* Seeman, T.E., L.F. Dubin, and M. Seeman, *Religiosity/spirituality and health. A critical review of the evidence for biological pathways.* Am Psychol, 2003. **58**(1): p. 53-63.

Koenig, H., D. King, and V.B. Carson, *Handbook of Religion and Health.* 2012: Oxford University Press.

Ano, G. and E. Vasconcelles, *Religious coping and psychological adjustment to stress: A meta-analysis.* J Clin Psychol, 2005. **61**: p. 461-80.

Ellison, C.G., et al., *Religious Involvement, Stress, and Mental Health: Findings from the 1995 Detroit Area Study*.* Social Forces, 2001. **80**(1): p. 215-249.

Strawbridge, W.J., et al., *Religious attendance increases survival by improving and maintaining good health behaviors, mental health, and social relationships.* Ann Behav Med, 2001. **23**(1): p. 68-74.

Van Ness, P.H., S.V. Kasl, and B.A. Jones, *Religion, race, and breast cancer survival.* Int J Psychiatry Med, 2003. **33**(4): p. 357-75.

83   **... many others stress the added role of taking part in organised services which are further enhanced by social and cultural factors:** Ferraro, K.F. and S. Kim, *Health benefits of religion among Black and White older adults? Race, religiosity, and C-reactive protein.* Soc Sci Med, 2014. **120**: p. 92-9.

Krause, N., *Church-based social support and health in old age: exploring variations by race.* J Gerontol B Psychol Sci Soc Sci, 2002. **57**(6): p. S332-47.

Debnam, K., et al., *Relationship between religious social support and general social support with health behaviors in a national sample of African Americans.* J Behav Med, 2012. **35**(2): p. 179-89.

Chida, Y., A. Steptoe, and L.H. Powell, *Religiosity/spirituality and mortality.*

83   **Religious practice is also a coping mechanism:** Ano, G. and E. Vasconcelles, *Religious coping and psychological adjustment to stress.*

83   **Although the association between religion and mental health issues ... is complex, the overall association between religion and mental health is positive:** Hackney, C.H. and G.S. Sanders, *Religiosity and Mental Health: A Meta-Analysis of Recent Studies.* J Sci Study Relig, 2003. **42**(1): p. 43-55.

Deaton, A. and A.A. Stone, *Two happiness puzzles.* Am Econ Rev, 2013. **103**(3): p. 591-597.

Myers, D.G. and E. Diener, *The Scientific Pursuit of Happiness.* 2018. **13**(2): p. 218-225.

83   **In countries where the state is responsible for important aspects of quality of life ... religion is not such a strong predictor of life satisfaction:** Zuckerman, M., C. Li, and E. Diener, *Religion as an Exchange System: The Interchangeability of God and Government in a Provider Role.* Pers Soc Psychol Bull, 2018. **44**(8): p. 1201-1213.

83   **This suggests that religion is, at least in part, a means for fulfilling certain needs:** Graham, C. and Crown, S., *Religion and well-being around the world: Social purpose, social time, or social insurance?* Int J Wellbeing, 2014. **4**(1).

Diener, E. and M.Y. Chan, *Happy people live longer: Subjective well-being contributes to health and longevity.* Appl Psychol: Health Well-Being, 2011. **3**(1): p. 1-43.

Tay, L., et al., *Religiosity and Subjective Well-Being: An International Perspective,* in *Religion and Spirituality Across Cultures,* C. Kim-Prieto, Editor. 2014, Springer Netherlands: Dordrecht. p. 163-175.

Diener, E., et al., *Advances and open questions in the science of subjective well-being.* Collabra: Psychology, 2018. **4**(1).

Koenig, H., D. King, and V.B. Carson, *Handbook of Religion and Health.*

83   **... in people with heart disease from birth, religious faith was positively associated with better quality of life:** Moons, P. and K. Luyckx, *Quality-of-life research in adult patients with congenital heart disease: current status and the way forward.* Acta Paediatr, 2019. **108**(10): p. 1765-1772.

83   **In people on dialysis for severe kidney disease and people with heart failure and recovering from heart attacks, it was also found to improve quality of life:** Burlacu, A., et al., *Religiosity, spirituality and quality of life of dialysis patients: a systematic review.* Int Urol Nephrol, 2019. **51**(5): p. 839-850.

Abu, H.O., et al., *Association of religiosity and spirituality with quality of life in patients with cardiovascular disease: a systematic review.* Qual Life Res, 2018. **27**(11): p. 2777-2797.

83   … it is evident that laughter and purpose are core to longevity and good health: Eger, R.J. and Maridal J.H., *A statistical meta-analysis of the wellbeing literature.* Int J Wellbeing, 2015. **5**(2).

Diener, E. and M.Y. Chan, Happy people live longer.

CHAPTER 5

85   I will explain why this is the case and provide some solutions to improve sleep: Siegel, J.M., *Clues to the functions of mammalian sleep.* Nature, 2005. **437**(7063): p. 1264-71.

Porkka-Heiskanen, T., *Adenosine in sleep and wakefulness.* Ann Med, 1999. **31**(2): p. 125-9.

Frank, M.G., *The mystery of sleep function: current perspectives and future directions.* Rev Neurosci, 2006. **17**(4): p. 375-92.

University of California - Berkeley. *Stressed to the max? Deep sleep can rewire the anxious brain.* [2019 4 November 2019 June 12, 2020]; Available from: https://www.sciencedaily.com/releases/2019/11/191104124140.htm.

89   REM sleep disorder is more common with age and present in 10 per cent of people over 70: Molano J, Boeve B, and Roberts R et al, *Frequency of sleep disorders in community-dwelling elderly: The Mayo Clinic Study of Aging.* Neurology., 2009. **72**(Suppl 3:A107).

89   … one in ten of us will experience sleepwalking at some stage …. Fortunately, it is not associated with any significant underlying health problems: Stallman, H.M. and M. Kohler, *Prevalence of Sleepwalking: A Systematic Review and Meta-Analysis.* PlOS One, 2016. **11**(11): p. e0164769-e0164769.

90   Night terrors occur during deep sleep, don't require specific treatments: Llorente, M.D., et al., *Night terrors in adults: Phenomenology and relationship to psychopathology.* J Clin Psychiatry, 1992. **53**(11): p. 392-394.

90   Almost two thirds of us experience sleep paralysis at some time: Dahlitz, M. and J.D. Parkes, *Sleep paralysis.* Lancet, 1993. **341**(8842): p. 406-7.

91   … one in four of us experience hallucinations associated with stress or fatigue: Ohayon, M.M., *Prevalence of hallucinations and their pathological associations in the general population.* Psychiatry Res. 2000. **97**(2): p. 153-164.

91   Afternoon naptime typically coincides with a brief lag in the body's internal alerting signal … giving sleep an advantage and edge over the wakefulness drive: Division of Sleep Medicine Harvard Medical School. *Homeostatic sleep drive.* Healthy Sleep Web Site. [2008 June 9, 2020.]; Available from: http://healthysleep.med.harvard.edu/healthy/glossary/g-j#homeostatic-sleep-drive

92   **If a person has insomnia, then afternoon napping may confuse the body clock:** Clark, N. *How to power nap like a pro.* [2018 Nov 16, 2018 June 9, 2020.]; Available from: https://www.sleepcycle.com/how-to-fall-asleep/how-to-power-nap-like-a-pro/

92   **As people get older, they have more fragmented sleep and this is not uncommonly associated with napping ... It's best to work out one's own napping preference and stick with that:** Goldman, S.E., et al., *Association between nighttime sleep and napping in older adults.* Sleep, 2008. **31**(5): p. 733-40.

Leng, Y., et al., *Who Take Naps? Self-Reported and Objectively Measured Napping in Very Old Women.* The Journals of Gerontology. Series A, Biological sciences and medical sciences, 2018. **73**(3): p. 374-379.

Ben-Simon, E., et al., *Overanxious and underslept.* Nat Hum Behav, 2020. **4**: p. 100-110.

Division of Sleep Medicine at Harvard Medical School. *Why Sleep Matters. Benefits of Sleep.* [Healthy Sleep 2008 June 9, 2020.]; Available from: http://healthysleep.med.harvard.edu/healthy/media-index

Knoblauch, V., et al., *Age-related changes in the circadian modulation of sleep-spindle frequency during nap sleep.* Sleep, 2005. **28**(9): p. 1093-101.

Siegel, J.M., *Clues to the functions of mammalian sleep.* Nature, 2005. **437**(7063): p. 1264-71.

Porkka-Heiskanen, T., *Adenosine in sleep and wakefulness.* Ann Med, 1999. **31**(2): p. 125-9.

Frank, M.G., *The mystery of sleep function: current perspectives and future directions.* Rev Neurosci, 2006. **17**(4): p. 375-92.

Clark, N. *How to power nap like a pro.*

92   **Periods of sleep following learning consistently enhance our ability to retain the content:** Diekelmann, S. and J. Born, *The memory function of sleep.* Nat Rev Neurosci, 2010. **11**(2): p. 114-26.

92   **... while a full night of slumber stabilises emotions, a sleepless night triggers anxiety levels to rise by as much as 30 per cent:** Anwar, Y. *Stress to the max? Deep sleep can rewire the anxious brain.* [Mind & Body, Research 2019 November 4, 2019 July 31, 2020]; Available from: https://news.berkeley.edu/2019/11/04/deep-sleep-can-rewire-the-anxious-brain/

93   **... sleep is a natural, non-pharmaceutical remedy for anxiety:** Ben-Simon, E., et al., *Overanxious and underslept.*

93   **Even subtle nightly changes in sleep affect anxiety levels:** Chang, J., et al., *Circadian control of the secretory pathway maintains collagen homeostasis.* Nat Cell Biol, 2020. **22**(1): p. 74-86.

93   ... **what can stop us from getting enough NREM deep sleep?:** American Sleep Association (ASA). *Deep Sleep: How to get more of it.* [2019 11 June 2020]; Available from: https://www.sleepassociation.org/about-sleep/stages-of-sleep/deep-sleep/#Function_of_Deep_Sleep

93   **... it's best to take exercise earlier in the day rather than before bedtime. Some people find that eating a late evening meal leads to a troubled night's sleep, while others find it helps:** Adam, K., *Dietary Habits and Sleep After Bedtime Food Drinks.* Sleep, 1980. **3**(1): p. 47-58.

94   **Using sound stimulation, such as listening to pink or white noise, can enhance deep sleep:** Papalambros, N.A., et al., *Acoustic Enhancement of Sleep Slow Oscillations and Concomitant Memory Improvement in Older Adults.* Frontiers in Human Neurosci, 2017 Mar 8;11:109.

94   **... getting less than seven and more than nine hours of sleep after the age of 50 are both linked to future problems with mental abilities:** Scarlett, S., Kenny, R.A., et al., *Objective Sleep Duration in Older Adults: Results From The Irish Longitudinal Study on Ageing.* J Am Geriatr Soc, 2020. **68**(1): p. 120-128.

94   **... toxins that have accumulated during the day, including those implicated in dementia:** Eugene, A.R. and J. Masiak, *The Neuroprotective Aspects of Sleep.* MEDtube Sci, 2015. **3**(1): p. 35-40.

94   **It is important that these toxins and waste products are regularly cleared by the cerebrospinal fluid:** Baranello, R.J., et al., *Amyloid-beta protein clearance and degradation (ABCD) pathways and their role in Alzheimer's disease.* Curr Alzheimer Res, 2015. **12**(1): p. 32-46.

95   **... missing sleep for even one night was associated with higher tau levels ... Therefore, insomnia in mid-life should be treated with the same gravity as high blood pressure and diabetes:** Benedict, C., et al., *Effects of acute sleep loss on diurnal plasma dynamics of CNS health biomarkers in young men.* Neurology, 2020. **94**: (11) e1181-e1189.

Ooms, S., et al., *Effect of 1 night of total sleep deprivation on cerebrospinal fluid β amyloid 42 in healthy middle-aged men: a randomized clinical trial.* JAMA Neurol, 2014. **71**(8): p. 971-7.

Pandi-Perumal, S.R., et al., *Senescence, sleep, and circadian rhythms.* Ageing Res Rev, 2002. **1**(3): p. 559-604.

Della Monica, C., et al., *Rapid Eye Movement Sleep, Sleep Continuity and Slow Wave Sleep as Predictors of Cognition, Mood, and Subjective Sleep Quality in Healthy Men and Women, Aged 20-84 Years.* Front Psychiatry. 2018 Jun 22;9:255.

Fan, M., et al., *Sleep patterns, genetic susceptibility, and incident cardiovascular disease: a prospective study of 385 292 UK biobank participants.* Eur Heart J, 2020 Mar 14;41(11): p.1182-1189.

95   ... 'looking as fresh as a daisy' after a good night's sleep has fundamental biological roots: Chang, J., et al., *Circadian control of the secretory pathway maintains collagen homeostasis*

96   The reduction of oxygen to the heart can lead to heart attacks: Yaffe, K., et al., *Sleep-Disordered Breathing, Hypoxia, and Risk of Mild Cognitive Impairment and Dementia in Older Women.* JAMA, 2011. **306**(6): p. 613-619.

96   When oxygen drops, stress hormones surge ... contribute to high blood pressure: Osman, A.M., et al., *Obstructive sleep apnea: current perspectives.* Nat Sci Sleep, 2018. **10**: p. 21-34.

96   Sleep apnoea is present in 3 per cent of 20–44-year-olds, 11 per cent of 45–64-year-olds, rising to 20 per cent of people over 60: McMillan, A. and M.J. Morrell, *Sleep disordered breathing at the extremes of age: the elderly.* Breathe (Sheffield, England), 2016. **12**(1): p. 50-60.

Bixler, E.O., et al., *Effects of age on sleep apnea in men: I. Prevalence and severity.* Am J Respir Crit Care Med, 1998. **157**(1): p. 144-8.

97   Some cytokines also help to promote sleep: Olson, E.J. *Lack of sleep: Can it make you sick?* [2018 Nov 28, 2018 June 9, 2020]; Available from: https://www.mayoclinic.org/diseases-conditions/insomnia/expert-answers/lack-of-sleep/faq-20057757

97   Sleep deprivation reduces both the production and the release of protective cytokines: The Sleep Foundation. [2020 June 16, 2020]; Available from: https://www.sleepfoundation.org/

97   Sound sleep also improves the action of immune T cells on fighting infection ... Chronic poor sleepers get more colds and flus and even have a poorer response to vaccinations: Perras, B. and J. Born, *Sleep associated endocrine and immune changes in the elderly*, in *Advances in Cell Aging and Gerontology.* 2005, Elsevier. p. 113-154.

University of Washington Health Sciences/UW Medicine. *Chronic sleep deprivation suppresses immune system: Study one of first conducted outside of sleep lab.* [2017 January 27, 2017 June 9, 2020]; Available from: www.sciencedaily.com/releases/2017/01/170127113010.htm

Phillips, D.J., M.I. Savenkova, and I.N. Karatsoreos, *Environmental disruption of the circadian clock leads to altered sleep and immune responses in mouse.* Brain Behav Immun, 2015. **47**: p. 14-23.

Bryant, P.A., J. Trinder, and N. Curtis, *Sick and tired: Does sleep have a vital role in the immune system?* Nat Rev Immunol, 2004. **4**(6): p. 457-67.

Van Someren, E.J.W., *Circadian and sleep disturbances in the elderly.* Experimental Gerontology, 2000. **35**(9): p. 1229-1237.

Santos, R.V.T., et al., *Moderate exercise training modulates cytokine profile and sleep in elderly people.* Cytokine, 2012. **60**(3): p. 731-735.

Prinz, P.N., *Age impairments in sleep, metabolic and immune functions.* Exp Gerontol, 2004. **39**(11-12): p. 1739-43.

Wang, D., et al., *The effect of sleep duration and sleep quality on hypertension in middle-aged and older Chinese: the Dongfeng-Tongji Cohort Study.* Sleep Med, 2017. **40**: p. 78-83.

Shi, G., et al., *A Rare Mutation of -(1)-Adrenergic Receptor Affects Sleep/Wake Behaviors.* Neuron, 2019. **103**(6): p. 1044-1055 e7.

Olson, E.J. *Lack of sleep*: Can it make you sick? Mayo Clinic Website. Nov 28, 2018, June 9, 2020. Available from: https://www.mayoclinic.org/diseases-conditions/insomnia/expert-answers/lack-of-sleep/faq-20057757

98   **... at the same time and same rhythm, through a central control system located in the brain, called the suprachiasmic nucleus:** Morin, L.P. and C.N. Allen, *The circadian visual system, 2005.* Brain Res Rev, 2006. **51**(1): p. 1-60.

Reppert, S.M. and D.R. Weaver, *Coordination of circadian timing in mammals.* Nature, 2002. **418**(6901): p. 935-41.

99   **Ageing is closely linked to circadian rhythms:** Lin, J.B., K. Tsubota, and R.S. Apte, *A glimpse at the aging eye.* npj Aging and Mech Dis 2, 16003 (2016).

Lucas, R.J., et al., *Diminished pupillary light reflex at high irradiances in melanopsin-knockout mice.* Science, 2003. **299**(5604): p. 245-7.

Lucas, R.J., et al., *How rod, cone, and melanopsin photoreceptors come together to enlighten the mammalian circadian clock.* Prog Brain Res, 2012. **199**: p. 1-18.

100   **... skin and liver cells retain 24-hour circadian rhythm even after getting rid of this gene ... If we could manipulate these genes to be more efficient it would slow down cell ageing:** Ray, S., et al., *Circadian rhythms in the absence of the clock gene Bmal1.* Science, 2020. **367**(6479): p. 800-806.

100   **Melatonin is the hormone which regulates the sleep–wake cycle ... It is primarily released by the pineal gland in the brain in response to darkness:** Zisapel, N., *New perspectives on the role of melatonin in human sleep, circadian rhythms and their regulation.* Br J Pharmacol, 2018. **175**(16): p. 3190-3199.

Auld, F., et al., *Evidence for the efficacy of melatonin in the treatment of primary adult sleep disorders.* Sleep Med Rev, 2017 Aug;34: p.10-22.

Faraone, S., *ADHD: Non-Pharmacologic Interventions, An Issue of Child and Adolescent Psychiatric Clinics of North America.* 2014, Elsevier.

100   **Melatonin's actions are not confined to sleep regulation – it also has antioxidant properties**: Chattoraj, A., et al., *Melatonin formation in mammals: in vivo perspectives.* Rev Endocr Metab Disord, 2009. **10**(4): p. 237-43.

100   **Light stimulus blocks the production of melatonin:** Reiter, R.J., *Pineal melatonin: cell biology of its synthesis and of its physiological interactions.* Endocr Rev, 1991. **12**(2): p. 151-80.

100   **... as a consequence, during the daily light period its level is very low:** Dominguez-Rodriguez, A., P. Abreu-Gonzalez, and R.J. Reiter, *Clinical aspects of melatonin in the acute coronary syndrome.* Curr Vasc Pharmacol, 2009. **7**(3): p. 367-73.

Waldhauser, F., J. Kovács, and E. Reiter, *Age-related changes in melatonin levels in humans and its potential consequences for sleep disorders.* Exp Gerontol, 1998. **33**(7-8): p. 759-72.

100   **Melatonin production decreases with age:** Emet, M., et al., *A Review of Melatonin, Its Receptors and Drugs.* Eurasian J Med, 2016. **48**(2): p. 135-41.

... vision also declines and eye diseases such as cataracts become more common: Duggan, E., Kenny, R.A., et al., *Time to Refocus Assessment of Vision in Older Adults? Contrast Sensitivity but Not Visual Acuity Is Associated With Gait in Older Adults.* J Gerontol A Biol Sci Med Sci, 2017. **72**(12): p. 1663-1668.

Connolly, E., Kenny, R.A., et al., *Prevalence of age-related macular degeneration associated genetic risk factors and 4-year progression data in the Irish population.* Br J Ophthalmol, 2018. **102**(12): p. 1691-1695.

100   **These combine to reduce the intensity of the response of the eye to light:** Maynard, M.L., et al., *Intrinsically Photosensitive Retinal Ganglion Cell Function, Sleep Efficiency and Depression in Advanced Age-Related Macular Degeneration.* Invest Ophthalmol Vis Sci, 2017. **58**(2): p. 990-996.

Wulff, K. and R.G. Foster, *Insight into the Role of Photoreception and Light Intervention for Sleep and Neuropsychiatric Behaviour in the Elderly.* Curr Alzheimer Res, 2017. **14**(10): p. 1022-1029.

100   **... replacing the deficiency in this sleep-regulating hormone improves sleep:** Haimov, I., et al., *Sleep disorders and melatonin rhythms in elderly people.* BMJ, 1994. **309**(6948): 167.

Tordjman, S., et al., *Advances in the research of melatonin in autism spectrum disorders: literature review and new perspectives.* Int J Mol Sci, 2013. **14**(10): p. 20508-20542.

101   **'Slow release' melatonin tablets appear to be more efficient than faster acting melatonin ... has been approved for the short-term treatment of insomnia in people aged 55 years and over:** Wade, A.G., et al., *Prolonged release melatonin in the treatment of primary insomnia: evaluation of the age cut-off for short- and long-term response.* Curr Med Res Opin, 2011. **27**(1): p. 87-98.

Sateia, M.J., et al., *Clinical Practice Guideline for the Pharmacologic Treatment of Chronic Insomnia in Adults: An American Academy of Sleep Medicine Clinical Practice Guideline.* J Clin Sleep Med, 2017. **13**(2): p. 307-349.

Riemersma-van der Lek, R.F., et al., *Effect of bright light and melatonin on cognitive and noncognitive function in elderly residents of group care facilities: a randomized controlled trial.* JAMA, 2008. **299**(22): p. 2642-55.

101 **Melatonin is also used for the short-term treatment of sleeping problems from jet lag or shift work:** Matheson, E. and B.L. Hainer, *Insomnia: Pharmacologic Therapy.* Am Fam Physician, 2017. **96**(1): p. 29-35.

British National Formulary, *BNF 76.* 76 ed, ed. J.F. Committee. 2018: Pharmaceutical Press. 1640.

101 **... humans were predominantly exposed to, and their lives and evolution depended on, yellow light ... and blue light ...:** Scott, A.C., *Burning Planet. The Story of Fire Through Time.* 2018, UK: Oxford University Press. 256.

Scott, A.C., et al., *The interaction of fire and mankind: Introduction.* Philosophical Transactions of the Royal Society B: Biological Sciences, 2016. **371**(1696): p. 20150162.

101 **... the incandescent light bulb ... produced relatively little blue light:** Cornell University Program of Computer Graphics. *Light Source Spectra.* [2001 02/06/2001 June 10, 2020.]; Available from: http://www.graphics.cornell.edu/online/measurements/source-spectra/index.html

101 **The longer the exposure before sleep, the shorter the duration of sleep:** Hysing, M., et al., *Sleep and use of electronic devices in adolescence: results from a large population-based study.* BMJ Open, 2015. **5**(1): e006748.

102 **It is likely that the negative effects of blue light is exaggerated with age:** Kayumov, L., et al., *Blocking low-wavelength light prevents nocturnal melatonin suppression with no adverse effect on performance during simulated shift work.* J Clin Endocrinol Metab, 2005. **90**(5): p. 2755-61.

Burkhart, K. and J.R. Phelps, *Amber lenses to block blue light and improve sleep: a randomized trial.* Chronobiol Int, 2009. **26**(8): p. 1602-12.

Biello, S.M., et al., *Alterations in glutamatergic signaling contribute to the decline of circadian photoentrainment in aged mice.* Neurobiology of Aging, 2018. **66**: p. 75-84.

102 **This alignment to the circadian clock is called our 'chronotype':** Wright, K.P., et al., *Entrainment of the Human Circadian Clock to the Natural Light–Dark Cycle.* Current Biology, 2013. **23**(16): p. 1554-1558.

Rosenberg, J., et al., *"Early to bed, early to rise": Diffusion tensor imaging identifies chronotype-specificity.* NeuroImage, 2014. **84**: p. 428-434.

Geddes, L. *First physical evidence of why you're an owl or a lark.* [Health 2013 30 September 2013 June 12, 2020]; Available from: https://www.newscientist.com/article/dn24292-first-physical-evidence-of-why-youre-an-owl-or-a-lark/

103   **This gene is a member of the Period family of genes:** Matsumura, R. and M. Akashi, *Role of the clock gene Period3 in the human cell-autonomous circadian clock.* Genes Cells, 2019. **24**(2): p. 162-171.

Xu, Y., et al., *Association Between Period 3 Gene Polymorphisms and Adverse Effects of Antidepressants for Major Depressive Disorder.* Genet Test Mol Biomarkers, 2019. **23**(12): p. 843-849.

Leocadio-Miguel, M.A., et al., *PER3 gene regulation of sleep-wake behavior as a function of latitude.* Sleep Health, 2018. **4**(6): p. 572-578.

Cheng, P., et al., *Daytime Sleep Disturbance in Night Shift Work and the Role of PERIOD3.* J Clin Sleep Med, 2018. **14**(3): p. 393-400.

Golalipour, M., et al., *PER3 VNTR polymorphism in Multiple Sclerosis: A new insight to impact of sleep disturbances in MS.* Mult Scler Relat Disord, 2017. **17**: p. 84-86.

103   **... although chronotypes are hard-wired, they do change with age:** Didikoglu, A., et al., *Longitudinal change of sleep timing: association between chronotype and longevity in older adults.* Chronobiology International, 2019. **36**(9): p. 1285-1300.

105   **... lions are frequently high-achieving goal-setters:** Escribano, C. and J.F. Díaz-Morales, *Are achievement goals different among morning and evening-type adolescents?* Personality and Individual Differences, 2016. **88**: p. 57-61.

Hess, A. *10 highly successful people who wake up before 6 a.m.* [Careers 2018 17 May 2018 June 11, 2020]; Available from: https://www.cnbc.com/2018/05/17/10-highly-successful-people-who-wake-up-before-6-a-m.html.

105   **... wolves tend to be more creative:** Gjermunds, N., et al., *Musicians: Larks, Owls or Hummingbirds?* J Circardian Rhythms, 2019;17:4.

106   **If rats with 24-hour access to food are compared with rats who have 8-hour access:** Chaix, A., et al., *Time-Restricted Feeding Prevents Obesity and Metabolic Syndrome in Mice Lacking a Circadian Clock.* Cell Metabolism, 2019. **29**(2): p. 303-319.e4.

106   **For a pre-bed time night snack there are a host of sleep-promoting foods that enhance melatonin and neuropeptides:** Richard, D.M., et al., *L-Tryptophan: Basic Metabolic Functions, Behavioral Research and Therapeutic Indications.* Int J Tryptophan Res, 2009. **2**: p. 45-60.

St-Onge, M.-P., A. Mikic, and C.E. Pietrolungo, *Effects of Diet on Sleep Quality.* Advances in Nutrition, 2016. **7**(5): p. 938-949.

Halson, S.L., *Sleep in elite athletes and nutritional interventions to enhance sleep.* Sports medicine (Auckland, N.Z.), 2014. **44 Suppl 1**(Suppl 1): p. S13-S23.

106   **Others such as chamomile contain apigenin:** Zick, S.M., et al., *Preliminary examination of the efficacy and safety of a standardized chamomile extract for chronic*

primary insomnia: A randomized placebo-controlled pilot study. BMC Complementary and Alternative Medicine, 2011. **11**(1): p. 78.

106 **In a randomised control trial of 95 men:** Hansen, A.L., et al., *Fish consumption, sleep, daily functioning, and heart rate variability.* J Clin Sleep Med, 2014. **10**(5): p. 567-575.

106 **In another study of 1,848 people:** Yoneyama, S., et al., *Associations between rice, noodle, and bread intake and sleep quality in Japanese men and women.* PLoS One, 2014. **9**(8): p. e105198.

CHAPTER 6

108 **One UK study found that … young adults unlock their phone 85 times a day:** Andrews, S., et al., *Beyond Self-Report: Tools to Compare Estimated and Real-World Smartphone Use.* Plos One, 2015. **10**(10): p. e0139004.

108 **… where young adults were instructed not to use phones, they exhibited withdrawal symptoms:** Clayton, R.B., G. Leshner, and A. Almond, *The Extended iSelf: The Impact of iPhone Separation on Cognition, Emotion, and Physiology.* J Comput-Mediat Comm, 2015. **20**(2): p. 119-135.

108 **… relationships were evident between smartphone use and depression:** Harrison, G. and M. Lucassen. *Stress and anxiety in the digital age: The dark side of technology.* [2019 1 March 2019 July 21, 2020]; Available from: https://www.open.edu/openlearn/health-sports-psychology/mental-health/managing-stress-and-anxiety-the-digital-age-the-dark-side-technology

Elhai, J.D., et al., *Problematic smartphone use: A conceptual overview and systematic review of relations with anxiety and depression psychopathology.* J Affect Disord, 2017. **207**: p. 251-259.

108 **… internet use by older adults is much more moderate:** Lam, S.S.M., S. Jivraj, and S. Scholes, *Exploring the Relationship Between Internet Use and Mental Health Among Older Adults in England: Longitudinal Observational Study.* J Med Internet Res, 2020. **22**(7): p. e15683.

109 **… 'the quality of experience, produced through a person–environment transaction…':** Aldwin, C.M., *Stress, coping, and development: An integrative perspective, 2nd ed.* 2007, New York, NY, US: Guilford Press.

109 **Stress is identified either by our own feelings or by objective tests:** Li, A.W. and C.A. Goldsmith, *The effects of yoga on anxiety and stress.* Altern Med Rev, 2012. **17**(1): p. 21-35.

Juster, R.P., B.S. McEwen, and S.J. Lupien, *Allostatic load biomarkers of chronic stress and impact on health and cognition.* Neurosci Biobehav Rev, 2010. **35**(1): p. 2-16.

109 **… stress induces loss of pigmented hair with retention of non-pigmented:**
Tan, S. and R. Weller, *Sudden whitening of the hair in an 82-year-old woman: the 'overnight greying' phenomenon.* Clinical and experimental dermatology, 2012. **37**(4): p. 458.

110 **… an American dermatologist described a 63-year-old male patient:** Navarini, A.A., S. Nobbe, and R.M. Trüeb, *Marie Antoinette syndrome.* Arch Dermatol, 2009. **145**(6): p. 656.

110 **Senator John McCain … biographer describes how, as a prisoner of war in Vietnam, McCain's hair rapidly turned white:** Coram, R., *American Patriot: The Life and Wars of Coloney Bud Day.* 2007, US: Little, Brown and Company. 417.

Rochester, S.I. and F.T. Kiley, *Honor Bound: American Prisoners of war in Southeast Asia, 1961-1973.* 1999, US: Naval Inst Pr. 706.

110 **Harvard researchers have shed further light on how stress causes such rapid greying … revert the adverse impact of stress and possibly of accelerated ageing**: Zhang, B., et al., *Hyperactivation of sympathetic nerves drives depletion of melanocyte stem cells.* Nature, 2020. **577**(7792): p. 676-681.

111 **In a large Gallup poll from 140 countries:** GALLUP, *Gallup 2019 Global Emotions Report.* 2019: gallup.com.

111 **… the University of Southern California similarly showed that ratings of daily perceived stressfulness yield a paradox**: Stone, A.A., S. Schneider, and J.E. Broderick, *Psychological stress declines rapidly from age 50 in the United States: Yet another well-being paradox.* J Psychosom Res, 2017. **103**: p. 22-28.

111 **This nicely aligns with our research on life satisfaction and happiness, which similarly shows a U-shaped curve:** Ward, M., C.A. McGarrigle, and R.A. Kenny, *More than health: quality of life trajectories among older adults-findings from The Irish Longitudinal Study of Ageing (TILDA).* Qual Life Res, 2019. **28**(2): p. 429-439.

112 **Older persons are wiser, more likely to live in the present:** Horovitz, B. *The Secrets to Happiness as You Age.* [2017 September 6, 2017 July 21, 2020]; Available from: https://www.nextavenue.org/the-secret-to-chronic-happiness-as-you-age/

112 **We become more adept at dealing with stressful challenges as we age:** Antczak, S. *Does Wisdom Come With Age?* [Living 2018 April 30, 2018 July 21, 2020]; Available from: https://www.nextavenue.org/wisdom-come-age/

112 **Brain-imaging supports a biological explanation for wisdom:** Meeks, T.W. and D.V. Jeste, *Neurobiology of wisdom: a literature overview.* Arch Gen Psychiatry, 2009. **66**(4): p. 355-365.

112 **Imparting wisdom, through inter-generational sharing:** Jeste, D.V., et al., *Age-Friendly Communities Initiative: Public Health Approach to Promoting Successful Aging.* Am J Geriatr Psychiatry, 2016. **24**(12): p. 1158-1170.

112   **Teresa Seamen is the senior researcher behind a new and innovative experimental programme of mentoring:** Gen2Gen. *Generation to Generation.* [2020 August 4, 2020]; Available from: https://www.facebook.com/pg/iamGen2Gen/community/

113   **Whereas people in the Blue Zones do experience stress, they have developed buffering techniques ... live longer lives:** Buettner, D., *The Blue Zones. Lessons for living longer from the people who've lived the longest.* First Paperbacked. ed. 2009, Washington DC: National Geographic.

114   **Cortisol levels were also measured before, during and after each participant's speech:** Townsend, S.S.M., H.S. Kim, and B. Mesquita, *Are You Feeling What I'm Feeling? Emotional Similarity Buffers Stress.* Social Psychological and Personality Science, 2014. **5**(5): p. 526-533.

114   **Spending time in nature reduces stress and results in a feeling of being in control:** Gonzalez, M.T., et al., *Therapeutic horticulture in clinical depression: a prospective study.* Res Theory Nurs Pract, 2009. **23**(4): p. 312-28.

115   **Numerous studies show that gardening improves physical and emotional well-being:** Genter, C., et al., *The contribution of allotment gardening to health and wellbeing: A systematic review of the literature.* Br J Occup Ther, 2015. **78**(10): p. 593-605.

       Soga, M., K.J. Gaston, and Y. Yamaura, *Gardening is beneficial for health: A meta-analysis.* Prev Med Rep, 2016. **5**: p. 92-99.

115   **... gardening combines physical activity with social interaction and exposure to nature and sunlight:** Thompson, R., *Gardening for health: a regular dose of gardening.* Clin Med (Lond), 2018. **18**(3): p. 201-205.

115   **Digging, raking and mowing are particularly calorie intense:** Vaz, M., et al., *A compilation of energy costs of physical activities.* Public Health Nutr, 2005. **8**(7a): p. 1153-83.

115   **... the social benefits of such projects may delay the symptoms of dementia:** Simons, L.A., et al., *Lifestyle factors and risk of dementia: Dubbo Study of the elderly.* Med J Aust, 2006. **184**(2): p. 68-70.

115   **Patients who are recovering from a heart attack or stroke find that exercise in a garden is more effective:** Wolf, S.L., et al., *Effect of constraint-induced movement therapy on upper extremity function 3 to 9 months after stroke: the EXCITE randomized clinical trial.* JAMA, 2006. **296**(17): p. 2095-104.

115   **The studies reported a significant positive effect for gardening in a wide range of health outcomes:** Soga, M., K.J. Gaston, and Y. Yamaura, *Gardening is beneficial for health.*

116   ... the stress-relieving effects of gardening were tested by giving allotment gardeners a psychological task: Van Den Berg, A.E. and M.H.G. Custers, *Gardening Promotes Neuroendocrine and Affective Restoration from Stress.* J Health Psychol, 2011. **16**(1): p. 3-11.

116   ... changes in the severity of depression and the ability to concentrate were measured: Gonzalez, M.T., et al., *Therapeutic horticulture in clinical depression*

116   ... the evidence strongly supports engagement with gardening as a means of destressing: Van Den Berg, A.E. and M.H.G. Custers, *Gardening Promotes Neuroendocrine and Affective Restoration from Stress.*

116   The bacteria, Mycobacterium vaccae, triggers release of serotonin: Reber, S.O., et al., *Immunization with a heat-killed preparation of the environmental bacterium – Mycobacterium vaccae – promotes stress resilience in mice.* Proc Natl Acad Sci USA, 2016. **113**(22): p. E3130-E3139.

117   Being surrounded by greenery makes life more manageable: van Dillen, S.M., et al., *Greenspace in urban neighbourhoods and residents' health: adding quality to quantity.* J Epidemiol Community Health, 2012. **66**(6): e8.

117   Given that the natural environment can enhance mental and physical health, governments are starting to regenerate urban environments ... less stressed: Frumkin, H., *Beyond toxicity: human health and the natural environment.* Am J Prev Med, 2001. **20**(3): p. 234-40.

Kinzler, D. *Reduce pandemic stress and anxiety with gardening and greenery.* [Home and Garden 2020 Mar 21st 2020 July 22, 2020]; Available from: https://www.wctrib.com/lifestyle/home-and-garden/5005515-Reduce-pandemic-stress-and-anxiety-with-gardening-and-greenery

117   The Japanese Ministry of Agriculture, Forestry and Fisheries coined the term *shinrin-yoku:* Kaplan S and Talbot JF, *Psychological Benefits of a Wilderness Experience,* in *Behavior and the Natural Environment. Human Behavior and Environment (Advances in Theory and Research), vol 6.,* Altman I and Wohlwill JF, Editors. 1983, Springer, Boston, MA.

118   Growing research in the potential of the natural environment to enhance health and well-being accentuates the underuse of this resource as a health promotion tool: Park, B.J., et al., *The physiological effects of Shinrin-yoku (taking in the forest atmosphere or forest bathing): evidence from field experiments in 24 forests across Japan.* Environmental health and preventive medicine, 2010. **15**(1): p. 18-26.

Nielsen, A. and K. Nilsson, *Urban forestry for human health and wellbeing.* Urban Forestry & Urban Greening - Urban for Urban Green, 2007. **6**: p. 195-197.

Coley, R.L., W.C. Sullivan, and F.E. Kuo, *Where Does Community Grow?:The Social Context Created by Nature in Urban Public Housing.* Environment and Behavior, 1997. **29**(4): p. 468-494.

Thompson, C.W., et al., *Enhancing Health Through Access to Nature: How Effective are Interventions in Woodlands in Deprived Urban Communities? A Quasi-experimental Study in Scotland, UK.* Sustainability, 2019. **11**(12): p. 3317-3317.

IUFRO, *International Union of Forest Research Organisations* [July 2021]; Available from: https://www.iufro.org/discover/organization/

118   **More than three-quarters of respondents said they wished their families shared more meals with them**: Conklin, A.I., et al., *Social relationships and healthful dietary behaviour: evidence from over-50s in the EPIC cohort, UK.* Soc Sci Med, 2014. **100**(100): p. 167-75.

118   **One in five people over 75 are lonely when eating alone**: Swerling, G. *A million elderly people skipping meals because they find eating alone too loney, charity reveals.* [2019 5 November 2019 August 4, 2020]; Available from: https://www.telegraph.co.uk/news/2019/11/05/million-elderly-people-skipping-meals-find-eating-alone-lonely/

Tani, Y., et al., *Eating alone and depression in older men and women by cohabitation status: The JAGES longitudinal survey.* Age Ageing, 2015. **44**(6): p. 1019-26.

118   **People spend more time eating when with others than when alone**: Hamrick, K. *Americans Spend an Average of 37 Minutes a Day Preparing and Serving Food and Cleaning Up.* [2016 November 07, 2016 August 4, 2020]; Available from: https://www.ers.usda.gov/amber-waves/2016/november/americans-spend-an-average-of-37-minutes-a-day-preparing-and-serving-food-and-cleaning-up/

118   **A majority of older persons said that family mealtimes were important occasions**: SeniorLiving.org. *Senior Living: The Risks of Eating Alone.* [2018 April 19, 2018 August 4, 2020]; Available from: https://www.seniorliving.org/health/eating-alone-risk/

118   **Nearly half of adults' meals are eaten in front of the computer, in the car, on the go**: Hartman Group. *Dinner: The American Mealtime Ritual's Last Stand.* [2018 February 12, 2018 July 22, 2020]; Available from: https://www.hartman-group.com/press-releases/1268781429/dinner-the-american-mealtime-rituals-last-stand

119   **Sharing meals develops social skills in children and adolescents**: Ball, K., et al., *Is healthy behavior contagious: associations of social norms with physical activity and healthy eating.* International Journal of Behavioral Nutrition and Physical Activity, 2010. **7**(1): p. 86.

Bevelander, K.E., D.J. Anschütz, and R.C.M.E. Engels, *Social norms in food intake among normal weight and overweight children.* Appetite, 2012. **58**(3): p. 864-872.

119   **Mealtimes are the perfect opportunity to share the invaluable wisdom that older adults have accumulated:** Mental Health Ireland. *Mealtimes.* [2021 13 May 2021]; Available from: https://www.mentalhealthireland.ie/a-to-z/m/

119   ... a best seller about the many benefits of walking on mood and brain function:
O'Mara, S., *In Praise of Walking*. 2019: Bodley Head.

121   If we become used to walking and then stop: Currey, M., *Daily Rituals: How
Artists Work*. 2013: Penguin Random House USA.

121   Researchers from Stanford showed how walking boosts creative inspiration ...
creative juices continued to flow when a person sat back down shortly after a
walk: Oppezzo, M. and D.L. Schwartz, *Give your ideas some legs: The positive effect
of walking on creative thinking*. Journal of Experimental Psychology: Learning,
Memory, and Cognition, 2014. **40**(4): p. 1142-1152.

121   Walking and creativity are both stress busters and positive mood enhancers:
Kardan, O., et al., *Is the preference of natural versus man-made scenes driven by bottom-
up processing of the visual features of nature?* Front Psychol, 2015. **6**: p. 471-471.

Kelly, P., et al., *Walking on sunshine: scoping review of the evidence for walking and
mental health*. Br J Sports Med, 2018. **52**(12): p. 800-806.

122   Rigorous scientific studies have confirmed the value of the ancient practice of
meditation: Pickut, B.A., et al., *Mindfulness based intervention in Parkinson's disease
leads to structural brain changes on MRI: a randomized controlled longitudinal trial*.
Clin Neurol Neurosurg, 2013. **115**(12): p. 2419-25.

Donley, S., et al., *Use and perceived effectiveness of complementary therapies in
Parkinson's disease*. Parkinsonism Relat Disord, 2019. **58**: p. 46-49.

122   Meditation increases brain blood flow ... Neurotrophins ... rise as a
consequence: Tang, Y.-Y., et al., *Short-term meditation increases blood flow in
anterior cingulate cortex and insula*. Front Psychol, 2015. **6**: p. 212.

122   ... mitochondria, present in every cell in the brain and body, produce 90 per cent
of the cell's energy: Black, D.S. and G.M. Slavich, *Mindfulness meditation and the
immune system: a systematic review of randomized controlled trials*. Ann N Y Acad Sci,
2016. **1373**(1): p. 13-24.

122   ... given these remarkable holistic benefits, having a go at mediation is surely a
'no brainer'!: Peng, C.K., et al., *Heart rate dynamics during three forms of meditation*.
Int J Cardiol, 2004. **95**(1): p. 19-27.

Sudsuang, R., V. Chentanez, and K. Veluvan, *Effect of Buddhist meditation on serum
cortisol and total protein levels, blood pressure, pulse rate, lung volume and reaction time*.
Physiol Behav, 1991. **50**(3): p. 543-8.

Wenneberg, S.R., et al., *A controlled study of the effects of the Transcendental
Meditation program on cardiovascular reactivity and ambulatory blood pressure*. Int J
Neurosci, 1997. **89**(1-2): p. 15-28.

122   'Life is available only in the present moment which underscores the principal
       behind mindfulness.':Thích Nhát Hanh, *Taming the Tiger Within: Meditations on
       Transforming Difficult Emotions.* 2004: Riverhead Books.

122   ... research shows that the ability to engage in this has many physical,
       psychological and cognitive benefits: Conklin, Q.A., et al., *Meditation, stress
       processes, and telomere biology.* Curr Opin Psychol, 2019. **28**: p. 92-101.

       Bower, J.E. and M.R. Irwin, *Mind-body therapies and control of inflammatory biology:
       A descriptive review.* Brain Behav Immun, 2016. **51**: p. 1-11.

123   Dispositional mindfulness is a quality in life: Tomasulo, D., *American Snake Pit:
       Hope, Grit, and Resilience in the Wake of Willowbrook.* 2018: Stillhouse Press. 290.

       Tomasulo, D., *Learned Hopefulness: The Power of Positivity to Overcome Depression.*
       2020: New Harbinger Publications. 192.

123   More trials are necessary to confirm these promising observations: Black, D.S.
       and G.M. Slavich, *Mindfulness meditation and the immune system.*

124   6 per cent of Americans being recommended yoga by a physician or therapist:
       Jeter, P.E., et al., *Yoga as a therapeutic intervention: a bibliometric analysis of published
       research studies from 1967 to 2013.* The Journal of Alternative and Complementary
       Medicine, 2015. **21**(10): p. 586-592.

124   ... yoga is promoted by the National Health Service ... combines physical
       postures, breathing techniques, relaxation and meditation: The Minded
       Institute. *Yoga in the NHS.* [2020 August 5, 2020]; Available from: https://
       themindedinstitute.com/yoga-in-healthcare/.

124   Studies on yoga have increased 50 fold since 2014: Jeter, P.E., et al., *Yoga as a
       therapeutic intervention.*

124   It works through a mix of increase in positive attitudes towards stress, self-
       awareness, coping mechanisms, control: Bonura, K.B., *The psychological benefits of
       yoga practice for older adults: Evidence and guidelines.* International Journal of Yoga
       Therapy, 2011. **21**(1): p. 129-142.

       Sherman, K.J., et al., *Mediators of yoga and stretching for chronic low back pain.*
       Evidence-based Complementary and Alternative Medicine, 2013. **2013**. 130818.
       doi:10.1155/2013/130818

       Brown, R.P. and P.L. Gerbarg, *Sudarshan Kriya Yogic breathing in the treatment of
       stress, anxiety, and depression: part II—clinical applications and guidelines.* J Altern
       Complement Med, 2005. **11**(4): p. 711-717.

124   spirituality: Moadel, A.B., et al., *Randomized controlled trial of yoga among a
       multiethnic sample of breast cancer patients: effects on quality of life.* Journal of Clinical
       Oncology, 2007. **25**(28): p. 4387-4395.

124   **compassion and mindfulness:** Brown, K.W. and R.M. Ryan, *The benefits of being present: mindfulness and its role in psychological well-being.* J Pers Soc Psychol, 2003. **84**(4): p. 822.

Chiesa, A. and A. Serretti, *Mindfulness-based stress reduction for stress management in healthy people: a review and meta-analysis.* J Altern Complement Med, 2009. **15**(5): p. 593-600.

Evans, S., et al., *Protocol for a randomized controlled study of Iyengar yoga for youth with irritable bowel syndrome.* Trials, 2011. **12**(1): p. 1-19.

124   **At a cellular level, it reduces inflammation:** Kiecolt-Glaser, J.K., et al., *Stress, inflammation, and yoga practice.* Psychosom Med, 2010. **72**(2): p. 113-121.

124   **Yoga increases cannabinoid and opiate levels and affects nervous activity:** Purdy, J., *Chronic physical illness: a psychophysiological approach for chronic physical illness.* YJBM. 2013. **86**(1): p. 15-28.

Ross, A. and S. Thomas, *The health benefits of yoga and exercise: a review of comparison studies.* J Altern Complement Med, 2010. **16**(1): p. 3-12.

Black, D.S., et al., *Yogic meditation reverses NF--B and IRF-related transcriptome dynamics in leukocytes of family dementia caregivers in a randomized controlled trial.* Psychoneuroendocrinology, 2013. **38**(3): p. 348-355.

124   **which release chemicals that relax blood vessels:** Prabhakaran, D. and A.M. Chandrasekaran, *Yoga for the prevention of cardiovascular disease.* Nat Rev Cardiol, 2020.

Wolff, M., et al., *Impact of a short home-based yoga programme on blood pressure in patients with hypertension: a randomized controlled trial in primary care.* J Hum Hypertens, 2016. **30**(10): p. 599-605.

Thiyagarajan, R., et al., *Additional benefit of yoga to standard lifestyle modification on blood pressure in prehypertensive subjects: a randomized controlled study.* Hypertens Res, 2015. **38**(1): p. 48-55.

124   **With ageing, telomeres shorten:** Kaszubowska, L., *Telomere shortening and ageing of the immune system.* J Physiol Pharmacol, 2008. **59**(Suppl 9): p. 169-186.

Hornsby, P.J., *Telomerase and the aging process.* Exp Gerontol, 2007. **42**(7): p. 575-81.

Blackburn, E.H., C.W. Greider, and J.W. Szostak, *Telomeres and telomerase: the path from maize, Tetrahymena and yeast to human cancer and aging.* Nat Med, 2006. **12**(10): p. 1133-1138.

124   **Telomerase is an important enzyme:** López-Otín, C., et al., *The hallmarks of aging.* Cell, 2013. **153**(6): p. 1194-1217.

Jacobs, T.L., et al., *Intensive meditation training, immune cell telomerase activity, and psychological mediators.* Psychoneuroendocrinology, 2011. **36**(5): p. 664-681.

124 **In a number of studies, yoga affected telomerase and telomere length:**
Lengacher, C.A., et al., *Influence of mindfulness-based stress reduction (MBSR) on telomerase activity in women with breast cancer (BC).* Biol Res Nurs, 2014. **16**(4): p. 438-47.

Lavretsky, H., et al., *A pilot study of yogic meditation for family dementia caregivers with depressive symptoms: effects on mental health, cognition, and telomerase activity.* Int J Geriatr Psychiatry, 2013. **28**(1): p. 57-65.

Krishna, B.H., et al., *Association of leukocyte telomere length with oxidative stress in yoga practitioners.* JCDR, 2015. **9**(3): p. CC01-CC3.

124 **... showed enhancement in telomerase:** Tolahunase, M., R. Sagar, and R. Dada, *Impact of Yoga and Meditation on Cellular Aging in Apparently Healthy Individuals: A Prospective, Open-Label Single-Arm Exploratory Study.* Oxid Med Cell Longev, 2017. **2017**: p. 7928981.

125 **Other important indicators of cell ageing that we previously discussed ... also change to a more youthful profile with yoga**: Kumar, S.B., et al., *Telomerase activity and cellular aging might be positively modified by a yoga-based lifestyle intervention.* J Altern Complement Med, 2015. **21**(6): p. 370-2.

Krishna, B.H., et al., *Association of leukocyte telomere length with oxidative stress in yoga practitioners.*

Tolahunase, M., R. Sagar, and R. Dada, *Impact of Yoga and Meditation on Cellular Aging in Apparently Healthy Individuals*

CHAPTER 7

127 **Emperors were obsessed with immortality ... responsible for the emperors' deaths as well as those of noblemen who perished in the bid to achieve perpetuity:**
Soth, A. *Elixirs of Immortal Life Were a Deadly Obsession. Ironically Enough.* [Cabinet of Curiosities 2018 December 28, 2018 March 31, 2020.]; Available from: https://daily.jstor.org/elixir-immortal-life-deadly-obsessions/

Pettit, H. *Mysterious "eternal life" potion discovered inside 2,000-year-old bronze pot in ancient Chinese tomb.* [2019 2019, March 4 March 31, 2020.]; Available from: https://www.thesun.ie/tech/3822766/elixir-of-immortality-found-in-ancient-chinese-tomb-reveals-deadly-quest-to-cheat-death-by-drinking-lethal-chemicals/

128 **The famous Chinese poet Po Chu-I spent hours bending over an alembic, stirring concoctions:** Yoke, H.P., G.T. Chye, and D. Parker, *Po Chü-i's Poems on Immortality.* Harv J Asiat Stud, 1974. **34**: p. 163-186.

129   The queen mole rat, with the assistance of her cohort of males, amazingly
maintains a consistent rate of fecundity … the recipe for today's formula for
the 'elixir of youth' may reside within this obscure small unattractive mammal:
Foster, K.R. and F.L. Ratnieks, *A new eusocial vertebrate?* Trends Ecol Evol, 2005.
20(7): p. 363-4.

Olshansky S. Jay, Perry. D., Miller Richard A, Butler Robert N. *In pursuit of the
longevity dividend. What should we be doing to prepare for the unprecedented aging of
humanity?* [2006 Feb 28, 2006 April 1, 2020.]; Available from: https://www.the-
scientist.com/uncategorized/the-longevity-dividend-47757

131   … some of the hormonal and cellular pathways that influence the rate of ageing
in lower organisms … Much of our knowledge of why human cells age derives
from observations in such lower species: van Heemst, D., *Insulin, IGF-1 and
longevity.* Aging Dis, 2010. **1**(2): p. 147-57.

Beyea, J.A., et al., *Growth hormone (GH) receptor knockout mice reveal actions of GH
in lung development.* Proteomics, 2006. **6**(1): p. 341-348.

133   … many mutations lead to death or impairment: de Boer, J., et al., *Premature aging
in mice deficient in DNA repair and transcription.* Science, 2002. **296**(5571): p. 1276-9.

134   … the cover of *Time* magazine in 2015: Carstensen, L., *The New Age of Much Older
Age, in Time.* 2015.

134   In 1900 life expectancy of females was 47. In 2010, it was 79: Bell, F. and M.
Miller, *Life Tables for the Unites States Social Security Area 1900-2100.* 2005, Social
Security Administration, Office of the Chief Actuary, SSA Pub. No. 11-11536.

137   The female hormone oestrogen has protective cardiovascular benefits:
Palmisano, B.T., L. Zhu, and J.M. Stafford, *Role of Estrogens in the Regulation of
Liver Lipid Metabolism.* Adv Exp Med Biol, 2017. **1043**: p. 227-256.

137   'In theory, if mortality rates did not increase as usual during ageing, humans
would live hundreds of years' … It may be an evolutionary development which
gives them a reproductive advantage: Finch, C.E., *Longevity, Senescence and the
Genome.* May 1994: The University of Chicago Press Books.

138   A more realistically achievable goal than negligible senescence: Olshansky S. Jay,
"Can we justify efforts to slow the rate of aging in humans?" in *Presentation before the
Annual meeting of the Gerontological Society of America.* 2003.

138   This target is chosen because the risk of death and most other negative attributes
of ageing tends to rise exponentially: Brody, J.A. and M.D. Grant, *Age- associated
diseases and conditions: Implications for decreasing late life morbidity.* Aging Clinical
and Experimental Research, 2001. **13**(2): p. 64-67.

138   **Such a seven-year delay would yield health and longevity benefits greater than what could be achieved with the elimination of cancer**: Olshansky, S.Jay, *Simultaneous/multiple cause-delay (SIMCAD): an epidemiological approach to projecting mortality.* J Gerontol, 1987. **42**(4): p. 358-65.

138   **... once achieved, this seven-year delay would yield equal health and longevity benefits for all subsequent generations:** Olshansky, S.Jay, L. Hayflick, and B.A. Carnes, *Position statement on human aging.* J Gerontol A Biol Sci Med Sci, 2002. **57**(8): p. B292-7.

138   **... friendship, stress relief, laughter, purpose, sleep, foods, physical activity and positive attitude do exactly that:** McCrory, C., Kenny R.A., et al., *The lasting legacy of childhood adversity for disease risk in later life.* Health Psychol, 2015. **34**(7): p. 687-96.

World Health Organization, *Global Health and Ageing.* 2011: NIH, US.

CHAPTER 8

141   **The Roman technique of bathing followed a somewhat standardised pattern**: Encyclopaedia Britannica Editors. *Thermae.* [1998 30 March 2011 April 30, 2020]; Available from: https://www.britannica.com/technology/thermae

142   **The use of water for therapeutic purposes is an ancient practice:** Gianfaldoni, S., et al., *History of the Baths and Thermal Medicine.* Open Access Maced J Med Sci, 2017. **5**(4): p. 566-568.

142   **These days, water therapy is used to treat musculoskeletal disorders:** Mooventhan, A. and L. Nivethitha, *Scientific evidence-based effects of hydrotherapy on various systems of the body.* N Am J Med Sci, 2014. **6**(5): p. 199-209.

142   **There is copious evidence for the health benefits of cold water on a number of systems:** Shevchuk, N.A., *Hydrotherapy as a possible neuroleptic and sedative treatment.* Med Hypotheses, 2008. **70**(2): p. 230-8.

142   **The cold water immersion provides a stimulus to our physiological systems which is related to the phenomenon of hormesis:** Leslie,. M., *How can we use moderate stresses to fortify humans and slow aging?* Sci Aging Knowledge Environ, 2005. **2005**(26): p. nf49.

142   **... exposure of cells to mild stress stimulates the synthesis of proteins ... the triggering of one recovery mechanism in the cell improves the functioning of other repair and recovery systems**: Shevchuk, N.A., *Adapted cold shower as a potential treatment for depression.* Medical Hypotheses, 2008. **70**(5): p. 995-1001.

142   **A cold shower or cold water immersion is a physiological stress:** Arumugam, T.V., et al., *Hormesis/preconditioning mechanisms, the nervous system and aging.* Ageing Res Rev, 2006. **5**(2): p. 165-78.

Fonager, J., et al., *Mild stress-induced stimulation of heat-shock protein synthesis and improved functional ability of human fibroblasts undergoing aging in vitro.* Exp Gerontol, 2002. **37**(10-11): p. 1223-8.

Leslie, M., *How can we use moderate stresses to fortify humans and slow aging?*

143 **… the number of cold receptors in the skin is up to ten times more than warm receptors:** Iggo, A. and B.J. Iggo, *Impulse coding in primate cutaneous thermoreceptors in dynamic thermal conditions.* J Physiol (Paris), 1971. **63**(3): p. 287-90.

Woodworth, R.S. and H. Schlosberg, *Experimental psychology [by] Robert S. Woodworth [and] Harold Schlosberg.* 1965, New York: Holt, Rinehart and Winston.

143 **On exposure of the skin to cold water, blood vessels contract and raise blood pressure:** Drummond, P.D., *Immersion of the hand in ice water releases adrenergic vasoconstrictor tone in the ipsilateral temple.* Auton Neurosci, 2006. **128**(1-2): p. 70-5.

Arumugam, T.V., et al., *Hormesis/preconditioning mechanisms.*

143 **One of these chemicals is noradrenaline:** Jansky, L., et al., *Change in sympathetic activity, cardiovascular functions and plasma hormone concentrations due to cold water immersion in men.* Eur J Appl Physiol Occup Physiol, 1996. **74**(1-2): p. 148-52.

143 **Cold exposure also releases noradrenaline in the major brain areas:** Schmidt, R.F., ed. *Fundamentals of Sensory Physiology.* 1978, Springer-Verlag, New York. 286.

Encyclopaedia Britannica Editors. *Brain.* [1998 March 21, 2020 May 01, 2020]; Available from: https://www.britannica.com/science/brain

143 **Almost all of our organs use noradrenaline … any stimulus which enhances its activity is important to 'ageing' physiology:** Edvinsson, L., et al., *Effect of exogenous noradrenaline on local cerebral blood flow after osmotic opening of the blood–brain barrier in the rat.* J Physiol, 1978. **274**: p. 149-156.

Jedema, H.P., et al., *Chronic cold exposure potentiates CRH-evoked increases in electrophysiologic activity of locus coeruleus neurons.* Biol Psychiatry, 2001. **49**(4): p. 351-9.

Jedema, H.P. and A.A. Grace, *Chronic exposure to cold stress alters electrophysiological properties of locus coeruleus neurons recorded in vitro.* Neuropsychopharmacology, 2003. **28**(1): p. 63-72.

Nisenbaum, L.K., et al., *Prior exposure to chronic stress results in enhanced synthesis and release of hippocampal norepinephrine in response to a novel stressor.* J Neurosci, 1991. **11**(5): p. 1478-84.

143 **… stimuli that excite the release of brain noradrenaline, such as cold water, may prevent dementia:** Robertson, I.H., *A noradrenergic theory of cognitive reserve: implications for Alzheimer's disease.* Neurobiol Aging, 2013. **34**(1): p. 298-308.

144   **Noradrenaline is one of the chemicals involved in the sympathetic nervous system:** Wikipedia contributors. *Sympathetic Nervous System*. [2003 15 April 2020 May 8, 2020]; Available from: https://en.wikipedia.org/wiki/Sympathetic_nervous_system

Encyclopaedia Britannica Editors. *Autonomic Nervous System*. 1998 Jan 11, 2019 May 01, 2020]; Available from: https://www.britannica.com/science/autonomic-nervous-system

144   **... control blood flow throughout the body, mostly through increasing noradrenaline release:** Nakamoto, M., *Responses of sympathetic nervous system to cold exposure in vibration syndrome subjects and age-matched healthy controls.* Int Arch Occup Environ Health, 1990. **62**(2): p. 177-81.

Shevchuk, N.A., *Adapted cold shower as a potential treatment for depression.*

Jansky, L., et al., *Change in sympathetic activity, cardiovascular functions and plasma hormone concentrations due to cold water immersion in men*

144   **Cold water exposure causes a four-fold rise in endorphins:** Vaswani, K.K., C.W. Richard, 3rd and G.A. Tejwani, *Cold swim stress-induced changes in the levels of opioid peptides in the rat CNS and peripheral tissues.* Pharmacol Biochem Behav, 1988. **29**(1): p. 163-8.

Suzuki, K., et al., *Responses of the hypothalamic-pituitary-adrenal axis and pain threshold changes in the orofacial region upon cold pressor stimulation in normal volunteers.* Arch Oral Biol, 2007. **52**(8): p. 797-802.

Mizoguchi, H., et al., *[Met5]enkephalin and delta2-opioid receptors in the spinal cord are involved in the cold water swimming-induced antinociception in the mouse.* Life Sci, 1997. **61**(7): p. PL81-6.

144   **... enhanced well-being and suppression of pain through stimulation of opioid receptors:** *Endorphins.*, in *The Columbia Encyclopedia* P. Lagasse, Goldman, L, Hobson, A, Norton, SR., 2000, Columbia University Press.

Encyclopaedia Britannica Editors. *Endorphin.* [1998 5 Jan 2012 May 01, 2020]; Available from: https://www.britannica.com/science/endorphin

144   **... any cold water swimmer or showerer ... will attest to having far fewer winter colds:** Brenner, I.K., et al., *Immune changes in humans during cold exposure: effects of prior heating and exercise.* J Appl Physiol (1985), 1999. **87**(2): p. 699-710.

Eglin, C.M. and M.J. Tipton, *Repeated cold showers as a method of habituating humans to the initial responses to cold water immersion.* Eur J Appl Physiol, 2005. **93**(5-6): p. 624-9.

Castellani, J.W., Brenner, I.K., and S.G. Rhind, *Cold exposure: human immune responses and intracellular cytokine expression.* Med Sci Sports Exerc, 2002. **34**(12): p. 2013-20.

Jansky, L., et al., *Immune system of cold-exposed and cold-adapted humans.* Eur J Appl Physiol Occup Physiol, 1996. **72**(5-6): p. 445-50.

Sramek, P., et al., *Human physiological responses to immersion into water of different temperatures.* Eur J Appl Physiol, 2000. **81**(5): p. 436-42.

145   **91 per cent expressed the will to continue the routine:** Buijze, G.A., et al., *The Effect of Cold Showering on Health and Work: A Randomized Controlled Trial.* PLoS One, 2016. **11**(9): p. e0161749.

145   **... good evidence to support associations between cold water swimming and decreases in tension, fatigue, improvement in mood and memory:** Knechtle, B., et al., *Cold Water Swimming-Benefits and Risks: A Narrative Review.* Int J Environ Res Public Health, 2020. **17**(23): 8984.

Huttunen, P., L. Kokko, and V. Ylijukuri, *Winter swimming improves general well-being.* Int J Circumpolar Health, 2004. **63**(2): p. 140-4.

145   **Sound evolutionary theories support the reasons why we find cold water exposure so invigorating ... cold water exposure is hard-wired:** McCullough, L. and S. Arora, *Diagnosis and treatment of hypothermia.* Am Fam Physician, 2004. **70**(12): p. 2325-32.

Encyclopaedia Britannica Editors. *Human Nervous System.* [1998 Apr 09, 2020 April 30, 2020]; Available from: https://www.britannica.com/science/human-nervous-system

Nutt, D.J., *The neuropharmacology of serotonin and noradrenaline in depression.* Int Clin Psychopharmacol, 2002. **17 Suppl 1**: p. S1-12.

Encyclopaedia Britannica Editors. *Hypothalamus.* [1998 Jan 10, 2019 May 01, 2019]; Available from: https://www.britannica.com/science/hypothalamus

Holloszy, J.O. and E.K. Smith, *Longevity of cold-exposed rats: a reevaluation of the "rate-of-living theory".* J Appl Physiol (1985), 1986. **61**(5): p. 1656-60.

Tikuisis, P., *Heat balance precedes stabilization of body temperatures during cold water immersion.* J Appl Physiol (1985), 2003. **95**(1): p. 89-96.

Mooventhan, A. and L. Nivethitha, *Scientific evidence-based effects of hydrotherapy on various systems of the body*

Arumugam, T.V., et al., *Hormesis/preconditioning mechanisms.*

Iggo, A. and B.J. Iggo, *Impulse coding in primate cutaneous thermoreceptors in dynamic thermal conditions.*

Woodworth, R.S. and H. Schlosberg, *Experimental psychology [by] Robert S. Woodworth [and] Harold Schlosberg.*

Drummond, P.D., *Immersion of the hand in ice water releases adrenergic vasoconstrictor tone in the ipsilateral temple.*

Jansky, L., et al., *Change in sympathetic activity, cardiovascular functions and plasma hormone concentrations due to cold water immersion in men.*

Edvinsson, L., et al., *Effect of exogenous noradrenaline on local cerebral blood flow after osmotic opening of the blood-brain barrier in the rat.*

Jedema, H.P., et al., *Chronic cold exposure potentiates CRH-evoked increases in electrophysiologic activity of locus coeruleus neurons.*

Nisenbaum, L.K., et al., *Prior exposure to chronic stress results in enhanced synthesis and release of hippocampal norepinephrine in response to a novel stressor.*

Wikipedia contributors. *Sympathetic Nervous System.*

Vaswani, K.K., C.W. Richard 3rd, and G.A. Tejwani, *Cold swim stress-induced changes in the levels of opioid peptides in the rat CNS and peripheral tissues.*

Suzuki, K., et al., *Responses of the hypothalamic-pituitary-adrenal axis and pain threshold changes in the orofacial region upon cold pressor stimulation in normal volunteers.*

Mizoguchi, H., et al., *[Met5]enkephalin and delta2-opioid receptors in the spinal cord are involved in the cold water swimming-induced antinociception in the mouse.*

Endorphins., in *The Columbia Encyclopedia* P. Lagasse, Goldman, L, Hobson, A, Norton, SR.,

Encyclopaedia Britannica Editors. *Endorphin.*

146 **The literature on the role of cold water in the treatment of depression is extensive and longstanding:** Shevchuk, N.A., *Adapted cold shower as a potential treatment for depression.*

147 **... a good illustration of how cold water swimming alleviated depression:** van Tulleken, C., et al., *Open water swimming as a treatment for major depressive disorder.* BMJ Case Reports, 2018. **2018**: bcr-2018-225007.

147 **The sympathetic surge could induce a heart attack if vessels to the heart are already narrow because of atherosclerosis or clots:** Imai, Y., et al., *Acute myocardial infarction induced by alternating exposure to heat in a sauna and rapid cooling in cold water.* Cardiology, 1998. **90**(4): p. 299-301.

Manolis, A.S., et al., *Winter Swimming: Body Hardening and Cardiorespiratory Protection Via Sustainable Acclimation.* Curr Sports Med Rep, 2019. **18**(11): p. 401-415.

Buijze, G.A., et al., *The Effect of Cold Showering on Health and Work: A Randomized Controlled Trial.*

147 … **brief whole-body exposure to cold water (15–23 °C) is safe:** Sramek, P., et al., *Human physiological responses to immersion into water of different temperatures.*

Holloszy, J.O. and E.K. Smith, *Longevity of cold-exposed rats.*

147 **The effect on core body temperature is so negligible that hypothermia is hardly ever a concern:** Doufas, A.G. and D.I. Sessler, *Physiology and clinical relevance of induced hypothermia.* Neurocrit Care, 2004. **1**(4): p. 489-98.

Tikuisis, P., *Heat balance precedes stabilization of body temperatures during cold water immersion.*

147 … **it is also helpful for a well-known skin disorder that becomes more common with age:** Dyhre-Petersen, N. and P. Gazerani, *Presence and characteristics of senile pruritus among Danish elderly living in nursing homes.* Future Sci OA, 2019. **5**(6): p. FSO399.

147 **Hot water showers and frequent hot water bathing aggravates or even causes asteatotic eczema:** Roy, A., et al., *Plasma norepinephrine responses to cold challenge in depressed patients and normal controls.* Psychiatry Res, 1987. **21**(2): p. 161-8.

Sramek, P., et al., *Human physiological responses to immersion into water of different temperatures.*

Holloszy, J.O. and E.K. Smith, *Longevity of cold-exposed rats.*

148 … **proximity to the sea is linked to better mood, less depression, less anxiety and better overall well-being … some studies suggest that it is particularly so as we get older:** Dempsey, S., et al., *Coastal blue space and depression in older adults.* Health Place, 2018. **54**: p. 110-117.

148 … **proximity to the sea can add four to seven years to lifespan:** Poulain, M., A. Herm, and G. Pes, *The Blue Zones: areas of exceptional longevity around the world.* Vienna Yearb Popul Res, 2013. **11**: p. 87-108.

148 … **research has shown that positive benefits on mood and well-being are more evident with the more visual exposure to the sea:** Volker, S. and T. Kistemann, *Reprint of: "I'm always entirely happy when I'm here!" Urban blue enhancing human health and well-being in Cologne and Dusseldorf, Germany.* Soc Sci Med, 2013. **91**: p. 141-52.

Mackerron, G. and S. Mourato, *Happiness is Greater in Natural Environments.* Global Environmental Change, 2013. **23**: p. 992–1000.

148 **some studies show that it is particularly evident as we get older:** Nutsford, D., et al., *Residential exposure to visible blue space (but not green space) associated with lower psychological distress in a capital city.* Health Place, 2016. **39**: p. 70-8.

Finlay, J., et al., *Therapeutic landscapes and wellbeing in later life: Impacts of blue and green spaces for older adults.* Health Place, 2015. **34**: p. 97-106.

148 **Proximity to the sea also increases the likelihood of engaging in physical activity**: Foley, R., *Swimming in Ireland: Immersions in therapeutic blue space.* Health Place, 2015. **35**: p. 218-25.

Foley, R., *Swimming as an accretive practice in healthy blue space.* Emot Space Socy, 2017. **22**. p. 43-51.

CHAPTER 9

152 **High calorie foods trigger dopamine release in 'pleasure centres' in the brain … The result is obesity and obesity-related diseases:** Grippo, R.M., et al., *Dopamine Signaling in the Suprachiasmatic Nucleus Enables Weight Gain Associated with Hedonic Feeding.* Curr Biol, 2020. **30**(2): p. 196-208 e8.

152 **Food consumption, diet, genes and pathways related to metabolism and cell energy production are the most important controllers for how our cells age:** Duggal, N.A., *Reversing the immune ageing clock: lifestyle modifications and pharmacological interventions.* Biogerontology, 2018. **19**(6): p. 481-496.

155 **… rats and pigeons are pretty much the same size and have the same basal metabolic rate but pigeons live seven times longer than rats:** Montgomery, M.K., A.J. Hulbert, and W.A. Buttemer, *The long life of birds: the rat-pigeon comparison revisited.* PLoS One, 2011. **6**(8): e24138.

155 **… we found that 70 per cent of people over 50 in Ireland are overweight or obese:** Leahy, S., Nolan, A., O'Connell, J., Kenny, R.A. *Obesity in an ageing society: implications for health, physical function and health service utilisation.* 2014. The Irish Longitudinal Study on Ageing (TILDA). https://www.doi.org/10.38018/TildaRe.2014-01

156 **The basal metabolic rate of overweight and obese people is higher than people of normal weight:** Liu, X., et al., *Resting heart rate and risk of metabolic syndrome in adults: a dose-response meta-analysis of observational studies.* Acta Diabetol, 2017. **54**(3): p. 223-235.

Zhang, S.Y., et al., *Overweight, resting heart rate and prediabetes/diabetes: A population-based prospective cohort study among Inner Mongolians in China.* Scientific Reports, 2016. **6**: 23939.

156 **Green tea, cabbage, berries, spinach, hot peppers and coffee are examples of foods that increase brown fat production:** Velickovic, K., et al., *Caffeine exposure induces browning features in adipose tissue in vitro and in vivo.* Scientific Reports, 2019. **9**: 9104.

156 **So, to all intents and purposes, brown fat is good fat:** Virtanen, K.A., et al., *Functional brown adipose tissue in healthy adults.* N Engl J Med, 2009. **360**(15): p. 1518-25.

157 **This may also be a further reason why cold water exposure, including cold showers, is beneficial:** Cohen, P. and B.M. Spiegelman, *Brown and Beige Fat: Molecular Parts of a Thermogenic Machine.* Diabetes, 2015. **64**(7): p. 2346-51.

157 **… we are yet to fully understand the sophisticated interactions between genetics, physiology and cognitive behaviour that regulate energy and body weight:** Lam, Y.Y. and E. Ravussin, *Analysis of energy metabolism in humans: A review of methodologies.* Mol Metab, 2016. **5**(11): p. 1057-1071.

158 **It has been shown to contain catechin which, in mice, slows down brain ageing and increases nerve circuits:** Unno, K., et al., *Green Tea Catechins Trigger Immediate-Early Genes in the Hippocampus and Prevent Cognitive Decline and Lifespan Shortening.* Molecules, 2020. **25**(7): 1484.

159 **In most of the Blue Zones, one to three small glasses of red wine are consumed per day:** Sass, C. *What Is the "Blue Zone" Diet? A Nutritionist Explains the Eating Plan That May Help You Live Longer and Healthier.* [2019 January 28, 2020 April 3, 2020]; Available from: https://www.health.com/nutrition/blue-zone-diet

159 **The Mediterranean diet is based on traditional foods:** Martínez-González, M.A., A. Gea, and M. Ruiz-Canela, *The Mediterranean Diet and Cardiovascular Health.* Circ Res, 2019. **124**(5): p. 779-798.

159 **A recent review paper summarised information on the diet taken from a series of studies:** Dinu, M., et al., *Mediterranean diet and multiple health outcomes: an umbrella review of meta-analyses of observational studies and randomised trials.* Eur J Clin Nutr, 2018. **72**(1): p. 30-43.

161 **In rhesus monkeys, after 20 years of reduced calorie intake … the fasting monkeys also live 30 per cent longer:** Dorling, J.L., C.K. Martin, and L.M. Redman, *Calorie restriction for enhanced longevity: The role of novel dietary strategies in the present obesogenic environment.* Ageing Res Rev, 2020 Dec;64: 101038.

164 **An interesting study in obese humans with early (pre) diabetes:** Sutton, E.F., et al., *Early time-restricted feeding improves insulin sensitivity, blood pressure, and oxidative stress even without weight loss in men with prediabetes.* Cell Metab, 2018. **27**(6): p. 1212-1221. e3.

164 **Some organisms become dormant during periods of food scarcity:** Calixto, A., *Life without Food and the Implications for Neurodegeneration.* Adv Genet, 2015. **92**: p. 53-74.

164 **… many of the health benefits of intermittent fasting are not simply the result of reduced free radical production or weight loss … caloric restriction reduces the chances of getting cancer:** Mattson, M.P., V.D. Longo, and M. Harvie, *Impact of intermittent fasting on health and disease processes.* Ageing Res Rev, 2017. **39**: p. 46-58.

165 **In a multi-centre UK study carried out in 2017:** Lean, M.E.J., et al., *Primary care-led weight management for remission of type 2 diabetes (DiRECT): an open-label, cluster-randomised trial.* The Lancet, 2018. **391**(10120): p. 541-551.

165   An excellent review in the *New England Journal of Medicine* summarises the current science and concludes that fasting is evolutionarily embedded ... In animals, fasting introduced at any stage in adult life shows all of the cellular benefits detailed above, even in very old animals: de Cabo, R. and M.P. Mattson, *Effects of Intermittent Fasting on Health, Aging, and Disease.* N Engl J Med, 2019. **381**(26): p. 2541-2551.

166   Several laboratory studies have demonstrated beneficial immune protection via the action of resveratrol on the SIRT1 gene: Lee, I.H., *Mechanisms and disease implications of sirtuin-mediated autophagic regulation.* Exp Mol Med, 2019. **51**(9): p. 1-11.

166   To get the dose used in research studies, consumption of up to 2,000mg of resveratrol a day is recommended: de la Lastra, C.A. and I. Villegas, *Resveratrol as an anti-inflammatory and anti-aging agent: mechanisms and clinical implications.* Mol Nutr Food Res, 2005. **49**(5): p. 405-30.

167   ... fisetin ... manipulates mTOR: Niedernhofer, L.J. and P.D. Robbins, *Senotherapeutics for healthy ageing.* Nat Rev Drug Discov, 2018. **17**(5): p. 377.

167   The action of metformin, a treatment for type 2 diabetes, also mimics caloric restriction: Glossmann, H.H. and O.M.D. Lutz, *Metformin and Aging: A Review.* Gerontology, 2019. **65**(6): p. 581-590.

167   ... recent clinical studies have reported an anti-inflammatory role for metformin and a beneficial effect for pathways involved in mouse models of arthritis: Son, H.-J., et al., *Metformin attenuates experimental autoimmune arthritis through reciprocal regulation of Th17/Treg balance and osteoclastogenesis.* Mediators Inflamm, 2014. **2014**: 973986.

       Martin-Montalvo, A., et al., *Metformin improves healthspan and lifespan in mice.* Nat Commun, 2013. **4**: 2192.

       Campbell, J.M., et al., *Metformin reduces all-cause mortality and diseases of ageing independent of its effect on diabetes control: A systematic review and meta-analysis.* Ageing Res Rev, 2017. **40**: p. 31-44.

       Saisho, Y., *Metformin and Inflammation: Its Potential Beyond Glucose-lowering Effect.* Endocr Metab Immune Disord Drug Targets, 2015. **15** (3):196-205.

       Samaras, K., et al., *SAT-LB115 Metformin-Use Is Associated With Slowed Cognitive Decline and Reduced Incident Dementia in Older Adults With Type 2 Diabetes Mellitus: The Sydney Memory and Ageing Study.* Diabetes Care, 2020 Nov:43(11):2691-2701.

168   ... those who followed the Japanese government's recommended dietary regime had a 15 per cent lower death rate: Kurotani, K., et al., *Quality of diet and mortality among Japanese men and women: Japan Public Health Center based prospective study.* BMJ, 2016. **352**: i1209.

169 **Around 98 per cent of Japanese children walk or cycle to school:** Mori, N., F. Armada, and D.C. Willcox, *Walking to school in Japan and childhood obesity prevention: new lessons from an old policy.* Am J Public Health, 2012. **102**(11): p. 2068-73.

170 **Japan's healthcare system is one of the best in the world:** Miller, L., Lu, W. *These Are the World's Healthiest Nations.* [2019 24 February 2019 Jan 2021]; Available from: https://www.bloomberg.com/news/articles/2019-02-24/spain-tops-italy-as-world-s-healthiest-nation-while-u-s-slips

171 **Omega-3 fatty acids are crucial for optimal body and brain function:** Ruxton, C., et al., *The health benefits of omega-3 polyunsaturated fatty acids: a review of the evidence.* J Hum Nutr Diet, 2007. **20**(3): p. 275-85.

171 **... fish is considered one of the most heart-healthy foods you can eat:** Link, R. *15 Incredibly Heart-Healthy Foods.* [Nutrition 2018 March 5, 2018 April 3, 2020]; Available from: https://www.healthline.com/nutrition/heart-healthy-foods

171 **... many big studies show that people who eat fish regularly have a lower risk of heart attacks, strokes and death from heart disease:** Djousse, L., et al., *Fish consumption, omega-3 fatty acids and risk of heart failure: a meta-analysis.* Clin Nutr, 2012. **31**(6): 846-53.

Zheng, J., et al., *Fish consumption and CHD mortality: an updated meta-analysis of seventeen cohort studies.* Public Health Nutr, 2012. **15**(4): p. 725-37.

Chowdhury, R., et al., *Association between fish consumption, long chain omega 3 fatty acids, and risk of cerebrovascular disease: systematic review and meta-analysis.* BMJ, 2012. **345**: e6698.

Buscemi, S., et al., *Habitual fish intake and clinically silent carotid atherosclerosis.* Nutr J, 2014. **13**: 2.

171 **... fish eaters were 13 per cent and vegetarians 22 per cent less likely to have a heart attack than meat eaters:** Tong, T.Y.N., et al., *Risks of ischaemic heart disease and stroke in meat eaters, fish eaters, and vegetarians over 18 years of follow-up: results from the prospective EPIC-Oxford study.* BMJ, 2019. **366**: l4897.

171 **Fish also benefits the immune system:** Mendivil, C.O., *Dietary Fish, Fish Nutrients, and Immune Function: A Review.* Front Nutr, 2021. **7**: 617652.

171 **... omega-3 fats in fish are especially important for the brain and the eye:** McCann, J.C. and B.N. Ames, *Is docosahexaenoic acid, an n-3 long-chain polyunsaturated fatty acid, required for development of normal brain function? An overview of evidence from cognitive and behavioral tests in humans and animals.* Am J Clin Nutr, 2005. **82**(2): p. 281-95.

171 **people who eat fish regularly have more grey matter:** Roques, S., et al., *Metabolomics and fish nutrition: a review in the context of sustainable feed development.* Rev Aquac, 2020. **12**(1): p. 261-282.

Raji, C.A., et al., *Regular fish consumption and age-related brain gray matter loss.* Am J Prev Med, 2014. **47**(4): p. 444-51.

172   **Regular fish eaters are less likely to become depressed:** Grosso, G., et al., *Omega-3 fatty acids and depression: scientific evidence and biological mechanisms.* Oxid Med Cell Longev, 2014. **2014**: 313570.

172   **... in patients who have been diagnosed with depression, omega-3 fatty acids and fish reduce symptoms:** Sarris, J., D. Mischoulon, and I. Schweitzer, *Omega-3 for bipolar disorder: meta-analyses of use in mania and bipolar depression.* J Clin Psychiatry, 2012. **73**(1): p. 81-6.

Peet, M. and D.F. Horrobin, *A dose-ranging study of the effects of ethyl-eicosapentaenoate in patients with ongoing depression despite apparently adequate treatment with standard drugs.* Arch Gen Psychiatry, 2002. **59**(10): p. 913-9.

Lin, P.Y. and K.P. Su, *A meta-analytic review of double-blind, placebo-controlled trials of antidepressant efficacy of omega-3 fatty acids.* J Clin Psychiatry, 2007. **68**(7): p. 1056-61.

Grosso, G., et al., *Omega-3 fatty acids and depression.*

172   **In patients who had self-harmed, those randomised to omega oil supplement for 12 weeks in addition to standard psychiatric care achieved substantial reductions in markers of suicidal behaviour:** Hallahan, B., et al., *Omega-3 fatty acid supplementation in patients with recurrent self-harm. Single-centre double-blind randomised controlled trial.* Br J Psychiatry, 2007. **190**: p. 118-22.

172   **Fish also benefits our sleep:** Leech, J. *10 Reasons Why Good Sleep Is Important.* Nutrition. [2020 February 24 April 3, 2020.]; Available from: https://www.healthline.com/nutrition/10-reasons-why-good-sleep-is-important

172   **... and energy by day:** Hansen, A.L., et al., *Fish consumption, sleep, daily functioning, and heart rate variability.* J Clin Sleep Med, 2014. **10**(5): p. 567-575.

172   **One recent large-scale evaluation of combined evidence from a number of research studies to examine the effects of red meat:** Johnston, B.C., et al., *Unprocessed Red Meat and Processed Meat Consumption: Dietary Guideline Recommendations From the Nutritional Recommendations (NutriRECS) Consortium.* Ann Intern Med, 2019; 171(10):756-764.

173   **In Ireland, 29 per cent of 18 to 30-year-olds and 1 in 5 people over 50 are vitamin D deficient ... It is very hard to get enough vitamin D if we live at high altitudes:** Laird, E., Kenny, R.A., et al., *Vitamin D deficiency is associated with inflammation in older Irish adults.* J Clin Endocrinol Metab, 2014. **99**(5): p. 1807-15.

Laird, E., Kenny, R.A., et al., *Vitamin D and bone health: potential mechanisms.* Nutrients, 2010. **2**(7): p. 693-724.

174 **Vitamin D is important to the body in many other ways:** Vanherwegen, A.S., C. Gysemans, and C. Mathieu, *Regulation of Immune Function by Vitamin D and Its Use in Diseases of Immunity.* Endocrinol Metab Clin North Am, 2017. **46**(4): p. 1061-1094.

Bacchetta, J., et al., *Antibacterial responses by peritoneal macrophages are enhanced following vitamin D supplementation.* PLoS One, 2014. **9**(12): e116530.

Sloka, S., et al., *Predominance of Th2 polarization by vitamin D through a STAT6-dependent mechanism.* J Neuroinflammation, 2011. **8**: 56.

174 **Our research supports a role for vitamin D in reducing the severity of Covid infections, including reducing deaths:** Rhodes, J.M., Kenny, R. A., et al., *Perspective: Vitamin D deficiency and COVID-19 severity – plausibly linked by latitude, ethnicity, impacts on cytokines, ACE2 and thrombosis.* J Intern Med, 2021.289(1):p. 97-115.

Rhodes, J., Kenny, R. A., et al., *COVID-19 mortality increases with northerly latitude after adjustment for age suggesting a link with ultraviolet and vitamin D.* BMJ Nutr Prev Health, 2020 Jun 14;3(1):118-120.

Rhodes, J.M., Kenny, R. A., et al., *Letter: low population mortality from COVID-19 in countries south of latitude 35° North supports vitamin D as a factor determining severity. Authors' reply.* Aliment Pharmacol Ther, 2020. **52**(2): p. 412-413.

174 **... governments in Ireland and UK now recommend vitamin D supplements for at risk Covid-19 cohorts:** Martineau, A.R., et al., *Vitamin D supplementation to prevent acute respiratory infections: individual participant data meta-analysis.* Health Technol Assess, 2019. **23**(2): p. 1-44.

174 **Vitamin D may also be of benefit in age related inflammation**: Ferrucci, L. and E. Fabbri, *Inflammageing: chronic inflammation in ageing, cardiovascular disease, and frailty.* Nat Rev Cardiol, 2018. **15**(9): p. 505-522.

Di Rosa, M., et al., *Vitamin D3: a helpful immuno-modulator.* Immunology, 2011. **134**(2): p. 123-39.

174 **For preventing the most serious effects of Covid-19, our research showed that intake of at least 800IU:** Huang, C., et al., *Clinical features of patients infected with 2019 novel coronavirus in Wuhan, China.* Lancet, 2020. **395**(10223): p. 497-506.

Xu, Z., et al., *Pathological findings of COVID-19 associated with acute respiratory distress syndrome.* Lancet Respir Med, 2020.

Rhodes, J.M., Kenny, R. A., et al., *Perspective: Vitamin D deficiency and COVID-19 severity.*

Rhodes, J., Kenny, R. A., et al., *COVID-19 mortality increases with northerly latitude after adjustment for age suggesting a link with ultraviolet and vitamin D.*

Rhodes, J.M., et al., *Letter: low population mortality from COVID-19 in countries south of latitude 35° North supports vitamin D as a factor determining severity. Authors' reply.*

174 **Free radicals are the toxic molecules formed naturally in the cell during energy production. They cause 'oxidative stress':** Christen, W.G., et al., *Vitamin E and age-related cataract in a randomized trial of women.* Ophthalmology, 2008. **115**(5): p. 822-829 e1.

Christen, W.G., et al., *Vitamin E and age-related macular degeneration in a randomized trial of women.* Ophthalmology, 2010. **117**(6): p. 1163-8.

Christen, W.G., et al., *Age-related cataract in a randomized trial of vitamins E and C in men.* Arch Ophthalmol, 2010. **128**(11): p. 1397-405.

174 **In the USA, antioxidant supplements account for a big chunk of the total intake:** National Center for Health Statistics (NCHS). *National Health and Nutrition Examination Survey US* [2009 14 August 2020 August 27, 2020]; Available from: https://www.cdc.gov/nchs/nhanes/index.htm

175 **In one study, which included almost 40,000 healthy women 45 years of age and older … Another large study found no benefit for vitamin C, vitamin E or beta-carotene supplements:** Mursu, J., et al., *Dietary supplements and mortality rate in older women: the Iowa Women's Health Study.* Arch Intern Med, 2011. **171**(18): p. 1625-1633.

Song, Y., et al., *Effects of vitamins C and E and beta-carotene on the risk of type 2 diabetes in women at high risk of cardiovascular disease: a randomized controlled trial.* Am J Clin Nutr, 2009. **90**(2): p. 429-37.

Lee, I.M., et al., *Vitamin E in the primary prevention of cardiovascular disease and cancer: the Women's Health Study: a randomized controlled trial.* JAMA, 2005. **294**(1): p. 56-65.

Cook, N.R., et al., *A randomized factorial trial of vitamins C and E and beta carotene in the secondary prevention of cardiovascular events in women: results from the Women's Antioxidant Cardiovascular Study.* Arch Intern Med, 2007. **167**(15): p. 1610-8.

175 **The Physicians' Health Study II, which included more than 14,000 male doctors aged 50 or older, found that neither vitamin E nor vitamin C supplements reduced the risk:** Gaziano, J.M., et al., *Vitamins E and C in the prevention of prostate and total cancer in men: the Physicians' Health Study II randomized controlled trial.* JAMA, 2009. **301**(1): p. 52-62.

Sesso, H.D., et al., *Vitamins E and C in the prevention of cardiovascular disease in men: the Physicians' Health Study II randomized controlled trial.* JAMA, 2008. **300**(18): p. 2123-33.

Sesso, H.D., et al., *Multivitamins in the Prevention of Cardiovascular Disease in Men: The Physicians' Health Study II Randomized Controlled Trial.* JAMA, 2012. **308**(17): p. 1751-1760.

175 ... **selenium and vitamin E supplements, taken alone or together, failed to prevent prostate cancer:** Lippman, S.M., et al., *Effect of selenium and vitamin E on risk of prostate cancer and other cancers: the Selenium and Vitamin E Cancer Prevention Trial (SELECT).* JAMA, 2009. **301**(1): p. 39-51.

Klein, E.A., et al., *Vitamin E and the risk of prostate cancer: the Selenium and Vitamin E Cancer Prevention Trial (SELECT).* JAMA, 2011. **306**(14): p. 1549-56.

175 ... **given that good diets contain antioxidants and prevent aforementioned diseases, why are antioxidant supplements not of the same benefit?:** Crowe, F.L., et al., *Fruit and vegetable intake and mortality from ischaemic heart disease: results from the European Prospective Investigation into Cancer and Nutrition (EPIC)-Heart study.* Eur Heart J, 2011. **32**(10): p. 1235-43.

Jerome-Morais, A., A.M. Diamond, and M.E. Wright, *Dietary supplements and human health: for better or for worse?* Mol Nutr Food Res, 2011. **55**(1): p. 122-35.

176 **Other posited reasons are that the relationship between free radicals and health is more complex than previously thought**: Halliwell, B., *The antioxidant paradox: less paradoxical now?* Br J Clin Pharmacol, 2013. **75**(3): p. 637-644.

176 ...**diets high in antioxidants have multiple health benefits but there is insufficient evidence for antioxidant supplements to replace a healthy diet**: Goodman, M., et al., *Clinical trials of antioxidants as cancer prevention agents: past, present, and future.* Free Radic Biol Med, 2011. **51**(5): p. 1068-84.

U.S. Food and Drug Administration. *What You Need To Know About Dietary Supplements.* [2017 29 November April 6, 2020.]; Available from: https://www.fda.gov/food/buy-store-serve-safe-food/what-you-need-know-about-dietary-supplements

Gaziano, J.M., et al., *Vitamins E and C in the prevention of prostate and total cancer in men*

Sesso, H.D., et al., *Vitamins E and C in the prevention of cardiovascular disease in men: the Physicians' Health Study II randomized controlled trial.*

Lippman, S.M., et al., *Effect of selenium and vitamin E on risk of prostate cancer and other cancers*

Klein, E.A., et al., *Vitamin E and the risk of prostate cancer: the Selenium and Vitamin E Cancer Prevention Trial (*

Crowe, F.L., et al., *Fruit and vegetable intake and mortality from ischaemic heart disease*

Jerome-Morais, A., A.M. Diamond, and M.E. Wright, *Dietary supplements and human health: for better or for worse?*

Halliwell, B., *The antioxidant paradox: less paradoxical now?*

177   **The relationship between our microbiome and the food we eat is complex and important:** Young, E., *I contain multitudes. The microbes within us and a grander view of life*. First U.S. edition. ed. 2016, New York, NY: Ecco, an imprint of HarperCollinsPublishers. 355.

Enders, G., *Gut: The inside story of our body's most underrated organ*. 2015, Germany: Greystone Books.

177   **The story starts with the Hadza tribe in Tanzania, East Africa, a hunter-gatherer tribe, living beside Lake Eyasi:** de Vrieze, J., *Gut Instinct*. Science, 2014. **343**(6168): p. 241-243.

Spector, T., *The Diet Myth: The Real Science Behind What We Eat*. 2015: W&N.

177   **Diversity in microbiome is a good thing:** Knight, R., *Follow Your Gut: How the Ecosystem in Your Gut Determines Your Health, Mood and More*. 2015: Simon & Schuster /TED.

Davis, N. *The human microbiome: why our microbes could be key to our health*. [2018 26 March April 6, 2020.]; Available from: https://www.theguardian.com/news/2018/mar/26/the-human-microbiome-why-our-microbes-could-be-key-to-our-health

178   **… a wealth of studies that have expanded the possible causal role of the microbiome:** Anderson, S.C., Cryan, J. F., Dinan, T., *The Psychobiotic Revolution. Mood, Food and the New Science of the Gut-Brain Connection*. 2019: National Geographic.

Sandhu, K.V., et al., *Feeding the microbiota-gut-brain axis: diet, microbiome, and neuropsychiatry*. Transl Res, 2017. **179**: p. 223-244.

Knight, R., *Follow Your Gut*.

178   **… we need diverse microbes and therefore a diverse diet to keep the microbes 'interested and stimulated':** Valdes, A.M., et al., *Role of the gut microbiota in nutrition and health*. BMJ, 2018. **361**: p. k2179.

Spector, T., *The Diet Myth*

179   **Foods high in polyphenols:** Saxelby, C. *Top 100 polyphenols. What are they and why are they important?* [Superfoods 2011 June 15, 2020]; Available from: https://foodwatch.com.au/blog/super-foods/item/top-100-polyphenols-what-are-they-and-why-are-they-important.html

Saxelby, C., *Nutrition for Life*. 2020: Hardie Grant Books. 192.

179   **There are specific microbiota associated with longer life:** Biagi, E., et al., *Gut Microbiota and Extreme Longevity*. Curr Biol, 2016. **26**(11): p. 1480-5.

Haran, J.P., et al., *The nursing home elder microbiome stability and associations with age, frailty, nutrition and physical location*. J Med Microbiol, 2018. **67**(1): p. 40-51.

179 **... the message is that long-lived fit, healthy people have very diverse microbiota:** Piggott, D.A. and S. Tuddenham, *The gut microbiome and frailty.* Translational Research, 2020. **221**: p. 23-43.

179 **Emulsifiers are found in all Western processed foods ... Likewise, artificial sweeteners, although 'safe', also produce toxic chemicals via microbes:** Chassaing, B., et al., *Dietary emulsifiers directly alter human microbiota composition and gene expression ex vivo potentiating intestinal inflammation.* Gut, 2017. **66**(8): p. 1414-1427.

Vo, T.D., B.S. Lynch, and A. Roberts, *Dietary Exposures to Common Emulsifiers and Their Impact on the Gut Microbiota: Is There a Cause for Concern?* Comprehensive Reviews in Food Science and Food Safety, 2019. **18**(1): p. 31-47.

181 **... there is little evidence for which prebiotics or probiotics people should consume and when it comes to probiotics, it isn't certain that the microbes will colonise your gut when they get there:** Tsai, Y.-L., et al., *Probiotics, prebiotics and amelioration of diseases.* J Biomed Sci, 2019. **26**(1): 3.

Quigley, E.M.M., *Prebiotics and Probiotics in Digestive Health.* Clin Gastroenterol Hepatol, 2019. **17**(2): p. 333-344.

181 **If you are taking antibiotics or have irritable bowel syndrome, there is evidence that probiotics help:** National Health Service (NHS). *Probiotics.* [2018 27 November 2018 June 15, 2020]; Available from: https://www.nhs.uk/conditions/probiotics/

182 **Ben Eiseman, a surgeon from Colorado, published a paper with his team describing the successful treatment by rectal faecal transplant of four critically ill people:** Eiseman, B., et al., *Fecal enema as an adjunct in the treatment of pseudomembranous enterocolitis.* Surgery, 1958. **44**(5): p. 854-9.

CHAPTER 10

183 **In her hallmark paper on a large series of older US adults:** Lindau, S.T., et al., *A study of sexuality and health among older adults in the United States.* N Engl J Med, 2007. **357**(8): p. 762-74.

184 **This hormone promotes extensive additional brain activity, including empathy and trust:** Quintana, D.S., et al., *Oxytocin pathway gene networks in the human brain.* Nat Commun, 2019. **10**(1): 668.

184 **... couples who worked on art projects together, rather than alone, boosted levels of the hormone:** Kosfeld, M., et al., *Oxytocin increases trust in humans.* Nature, 2005. **435**(7042): p. 673-676.

184   **People on oxytocin have been shown to be more willing to entrust someone with their money:** Mikolajczak, M., et al., *Oxytocin not only increases trust when money is at stake, but also when con-fidential information is in the balance.* Biological Psychology, 2010. **85**(1): p. 182-184.

184   **There's a common misconception that as people age, they lose interest in sex:** Smith, L., et al., *Sexual Activity is Associated with Greater Enjoyment of Life in Older Adults.* J Sex Med, 2019. **7**(1): p. 11-18.

184   **... older people remain sexually active and continue to attribute importance to sex:** Lee, D.M., et al., *Sexual Health and Well-being Among Older Men and Women in England: Findings from the English Longitudinal Study of Ageing.* Arch Sex Behav, 2016. **45**(1): p. 133-44.

Schick, V., et al., *Sexual behaviors, condom use, and sexual health of Americans over 50: implications for sexual health promotion for older adults.* J Sex Med, 2010. **7 Suppl 5**: p. 315-29.

Lindau, S.T. and N. Gavrilova, *Sex, health, and years of sexually active life gained due to good health: evidence from two US population based cross sectional surveys of ageing.* BMJ, 2010. **340**: c810.

Dunn, K.M., P.R. Croft, and G.I. Hackett, *Association of sexual problems with social, psychological, and physical problems in men and women: a cross sectional population survey.* J Epidemiol Community Health, 1999. **53**(3): p. 144-8.

Laumann, E.O., et al., *Sexual problems among women and men aged 40-80 y: prevalence and correlates identified in the Global Study of Sexual Attitudes and Behaviors.* Int J Impot Res, 2005. **17**(1): p. 39-57.

Lindau, S.T., et al., *A study of sexuality and health among older adults in the United States*

185   **... our TILDA research reported 80 per cent of couples with an average age 64 consider sex to be important:** Orr, J., Layte, R., and O'Leary, N. *Sexual Activity and Relationship Quality in Middle and Older Age: Findings From The Irish Longitudinal Study on Ageing (TILDA).* J Gerontol B Psychol Sci Soc Sci, 2019. **74**(2): p. 287-297.

185   **Older adults in England enjoy life more when they are sexually active:** Lee, D.M., et al., *Sexual Health and Well-being Among Older Men and Women in England.*

185   **Although being sexually active is largely dependent on having a spouse or cohabiting partner, this is not exclusive:** Orr J, McGarrigle C, and Kenny RA, *Sexual activity in the over 50s population in Ireland.* 2017, Trinity College Dublin: TILDA (The Irish Longitudinal Study on Ageing).

Orr, J., R. Layte, N. O'Leary, Kenny, R. A., *Sexual Activity and Relationship Quality in Middle and Older Age.*

184 **Lindau's work showed that the frequency of sexual activity in older persons is similar to that of adults aged 18–59:** Laumann, E.O., et al., *The Social Organization of Sexuality. Sexual Practices in the United States.* 1994: The University of Chicago Press Books. 750.

184 **Couples who are regularly sexually active and satisfied with their sex lives are more satisfied overall with life as a couple:** Byers, E.S., *Relationship satisfaction and sexual satisfaction: a longitudinal study of individuals in long-term relationships.* J Sex Res, 2005. **42**(2): p. 113-8.

Fisher, W.A., et al., *Individual and Partner Correlates of Sexual Satisfaction and Relationship Happiness in Midlife Couples: Dyadic Analysis of the International Survey of Relationships.* Arch Sex Behav, 2015. **44**(6): p. 1609-20.

184 **Men and women who are sexually active have better memory and concentration:** Wright, H. and R.A. Jenks, *Sex on the brain! Associations between sexual activity and cognitive function in older age.* Age Ageing, 2016. **45**(2): p. 313-7.

Maunder, L., D. Schoemaker, and J.C. Pruessner, *Frequency of Penile-Vaginal Intercourse is Associated with Verbal Recognition Performance in Adult Women.* Arch Sex Behav, 2017. **46**(2): p. 441-453.

185 **Sexual satisfaction and frequency are associated with better communication within couples:** Gillespie, B.J., *Sexual Synchronicity and Communication Among Partnered Older Adults.* J Sex Marital Ther, 2017. **43**(5): p. 441-455.

185 **Higher endorphins benefit the immune system:** Plein, L.M. and H.L. Rittner, *Opioids and the immune system - friend or foe.* Br J Pharmacol, 2018. **175**(14): p. 2717-2725.

186 **Masters and Johnson were the amazing pioneers of sexual studies who carried out groundbreaking observations of sexual activity and its biological consequences:** Brecher, E.M., The Journal of Sex Research, 1970. **6**(3): p. 247-250.

186 **More recent studies have used wearable measurement technologies to determine energy expenditure during sex:** Frappier, J., et al., *Energy Expenditure during Sexual Activity in Young Healthy Couples.* Plos One, 2013. **8**(10): e79342.

187 **In many cases, when it comes to older people and sex, doctors, nurses and others often put their heads in the sand:** Gott, M., S. Hinchliff, and E. Galena, *General practitioner attitudes to discussing sexual health issues with older people.* Soc Sci Med, 2004. **58**(11): p. 2093-103.

Malta, S., et al., *Do you talk to your older patients about sexual health? Health practitioners' knowledge of, and attitudes towards, management of sexual health among older Australians.* Aust J Gen Pract, 2018. **47**(11): p. 807-811.

187   **... most of the biological issues which complicate sex in later life are amenable to investigation and treatment:** Heiman, J.R., et al., *Sexual satisfaction and relationship happiness in midlife and older couples in five countries.* Arch Sex Behav, 2011. **40**(4): p. 741-53.

Ambler, D.R., E.J. Bieber, and M.P. Diamond, *Sexual function in elderly women: a review of current literature.* Rev Obstet Gynecol, 2012. **5**(1): p. 16-27.

Muller, B., et al., *Sexuality and affection among elderly German men and women in long-term relationships: results of a prospective population-based study.* PLoS One, 2014. **9**(11): p. e111404.

187   **The researchers speculated, very reasonably, that the benefits were because of release of oxytocin, dopamine and other endorphins:** Wright, H. and R.A. Jenks, *Sex on the brain!*

188   **Other research in both humans and animals in the last decade underscores that frequent sexuality might enhance performance on brain abilities:** Wright, H., R. Jenks, and N. Demeyere, *Frequent Sexual Activity Predicts Specific Cognitive Abilities in Older Adults.* J Gerontol B Psychol Sci Soc Sci, 2017. **74** (1):47-51.

188   **In addition to vaginal and oral sex, masturbation, kissing, petting and fondling are all associated with better memory function:** Wright, H., R.A. Jenks, and D.M. Lee, *Sexual Expression and Cognitive Function: Gender-Divergent Associations in Older Adults.* Arch Sex Behav, 2020. **49**(3): p. 941-951.

Maunder, L, D. Schoemaker, and J.C. Pruessner, *Frequency of Penile-Vaginal Intercourse is Associated with Verbal Recognition Performance in Adult Women*

188   **A 2010 study discovered a link between sexual activity and growth of new brain cells in male rats:** Leuner, B., E.R. Glasper, and E. Gould, *Sexual experience promotes adult neurogenesis in the hippocampus despite an initial elevation in stress hormones.* PLOS One, 2010. **5**(7): p. e11597.

189   **... further male rat studies found that daily sexual activity was not only associated with the formation of new brain cells but also with enhanced brain function ... the 'reward' aspect of intercourse may be a mechanism by which new brain cells form:** Glasper, E.R. and E. Gould, *Sexual experience restores age-related decline in adult neurogenesis and hippocampal function.* Hippocampus, 2013. **23**(4): p. 303-12.

189   **Stress and depression both blunt the formation of new brain cells:** Spalding, K.L., et al., *Dynamics of hippocampal neurogenesis in adult humans.* Cell, 2013. **153**(6): p. 1219-1227.

189   **... vaginal sex increases levels of serotonin and oxytocin:** Allen, M.S., *Sexual Activity and Cognitive Decline in Older Adults.* Arch Sex Behav, 2018. **47**(6): p. 1711-1719.

Wright, H. and R.A. Jenks, *Sex on the brain!*

189   … **sexuality does not disappear – it is just less apparent:** Yoquinto, L. *Sex Life Becomes More Satisfying for Women After 40.* [2013 May 30, 2013 April 8, 2020.]; Available from: https://www.livescience.com/36073-women-sex-life-age.html

189   **Urinary infections after intercourse also become more frequent:** Raz, R., *Urinary tract infection in postmenopausal women.* Korean J Urol, 2011. **52**(12): p. 801-8.

190   … **a study of single women reported attitudes to and experiences of unconventional sexual relationships in older women:** von Sydow, K., *Unconventional sexual relationships: data about German women ages 50 to 91 years.* Arch Sex Behav, 1995. **24**(3): p. 271-90.

190   **One large study in California looked at sexual activity and sexual satisfaction in 1,300 healthy women … some older women who have no intimate contact of any kind are perfectly happy without it:** Trompeter, S.E., R. Bettencourt, and E. Barrett-Connor, *Sexual activity and satisfaction in healthy community-dwelling older women.* Am J Med, 2012. **125**(1): p. 37-43 e1.

      Orr, J., R. Layte, N. O'Leary, Kenny, R. A., *Sexual Activity and Relationship Quality in Middle and Older Age.*

      Yoquinto, L. *Sex Life Becomes More Satisfying for Women After 40.*

191   **Eighty-five per cent of men in the UK aged 60–69 are sexually active:** Lindau, S.T., et al., *A study of sexuality and health among older adults in the United States*

192   **Many men experience ED during times of stress. It can also be a sign of emotional or relationship difficulties … testosterone therapy may also be effective if levels are low:** Schaefer, A. *12 Surprising Facts About Erections.* [2015 December 4, 2017 April 8, 2020]; Available from: https://www.healthline.com/health/erectile-dysfunction/surprising-facts#1

      Ferguson, S. *Everything You Need to Know About Penis Health.* [2019 March 26]; Available from: https://www.healthline.com/health/penis-health

      York, S., Nicholls, E. *All About the Male Sex Drive.* [2017 October 10, 2019. April 8, 2020]; Available from: https://www.healthline.com/health/mens-health/sex-drive

      Cheng, J.Y.W., et al., *Alcohol consumption and erectile dysfunction: meta-analysis of population-based studies.* Int J Impot Res, 2007. **19**(4): p. 343-352.

      Healthline Editorial Team. *A List of Blood Pressure Medications.* [2019 April 7, 2020. April 8, 2020]; Available from: https://www.healthline.com/health/high-blood-pressure-hypertension-medication

CHAPTER 11

195   **Jerry Morris and Margaret Crawford, observed that they appeared to be doing more post mortems on bus drivers than bus conductors … we had clear evidence**

that sedentary occupations were more likely to kill: Morris, J.N. and M.D. Crawford, *Coronary heart disease and physical activity of work; evidence of a national necropsy survey.* BMJ, 1958. **2**(5111): p. 1485-1496.

196   **... in one large analysis of almost a million people followed for 20 years, inactive people were 40 per cent more likely to die early:** Nocon, M., et al., *Association of physical activity with all-cause and cardiovascular mortality: a systematic review and meta-analysis.* Eur J Cardiovasc Prev Rehabil, 2008. **15**(3): p. 239-46.

197   **Regular physical activity also improves mental health and well-being:** Teychenne, M., K. Ball, and J. Salmon, *Physical activity and likelihood of depression in adults: a review.* Prev Med, 2008. **46**(5): p. 397-411.

197   **or alleviates depression:** Conn, V.S., *Depressive symptom outcomes of physical activity interventions: meta-analysis findings.* Ann Behav Med, 2010. **39**(2): p. 128-38.

197   **increases vitality and an optimistic approach ... BDNF is protective against stress, which partly explains why we often feel so at ease and in a happier state and issues and problems appear clearer after exercising:** Reed, J. and D. Ones, *The effect of acute aerobic exercise on positive activated affect: A meta-analysis.* Psychol Sport Exerc, 2006. **7**: p. 477-514.

Puetz, T.W., P.J. O'Connor, and R.K. Dishman, *Effects of chronic exercise on feelings of energy and fatigue: a quantitative synthesis.* Psychol Bull, 2006. **132**(6): p. 866-76.

198   **BDNF released during exercise supports the growth of new nerve cells:** Coelho, F.G.d.M., et al., *Physical exercise modulates peripheral levels of brain-derived neurotrophic factor (BDNF): A systematic review of experimental studies in the elderly.* Arch Gerontol Geriatr, 2013. **56**(1): p. 10-15.

Erickson, K.I., et al., *Exercise training increases size of hippocampus and improves memory.* Proc Natl Acad Sci USA, 2011. **108**(7): p. 3017-22.

198   **... a publication in the cruelly named** *American Journal of Insanity* **described the benefits of exercise as a treatment for depression:** Shepherd Ivory Franz, and G. V. Hamilton, *The effects of exercise upon the retardation in conditions of depression.* Am J Psychiatry, 1905. **62**(2): p. 239-256.

198   **... a number of chemicals released by the brain during exercise that are important both for prevention and treatment of depression and anxiety have been discovered:** Deslandes, A., et al., *Exercise and mental health: many reasons to move.* Neuropsychobiology, 2009. **59**(4): p. 191-8.

198   **Exercise confers additional psychological benefits:** Daley, A., *Exercise and depression: a review of reviews.* J Clin Psychol Med Settings, 2008. **15**(2): p. 140-7.

Martinsen, E.W., *Physical activity in the prevention and treatment of anxiety and depression.* Nord J Psychiatry, 2008. **62 Suppl 47**: p. 25-9.

198   **... research has also shown that physical activity levels are low in adults who suffer with depression:** López-Torres Hidalgo, J., et al., *Effectiveness of physical exercise in the treatment of depression in older adults as an alternative to antidepressant drugs in primary care.* BMC Psychiatry, 19, 21 (2019).

Hamer, M., K.L. Lavoie, and S.L. Bacon, *Taking up physical activity in later life and healthy ageing: the English longitudinal study of ageing.* Br J Sports Med, 2014. **48**(3): p. 239-43.

Mammen, G. and G. Faulkner, *Physical activity and the prevention of depression: a systematic review of prospective studies.* Am J Prev Med, 2013. **45**(5): p. 649-57.

Donoghue, O., M. O'Connell, and R.A. Kenny, *Walking to wellbeing: physical activity, social participation and psychological health in Irish adults aged 50 years and older.* Dublin: The Irish longitudinal study on ageing (TILDA), 2016.

Teychenne, M., K. Ball, and J. Salmon, *Physical activity and likelihood of depression in adults*

199   **Remarkably, exercise increases the size of the hippocampus ... The increase in hippocampus size also increases release of BDNF:** Hillman, C.H., K.I. Erickson, and A.F. Kramer, *Be smart, exercise your heart: exercise effects on brain and cognition.* Nat Rev Neurosci, 2008. **9**(1): p. 58-65.

van Praag, H., et al., *Exercise enhances learning and hippocampal neurogenesis in aged mice.* J Neurosci, 2005. **25**(38): p. 8680-5.

Cotman, C.W. and N.C. Berchtold, *Exercise: a behavioral intervention to enhance brain health and plasticity.* Trends Neurosci, 2002. **25**(6): p. 295-301.

Creer, D.J., et al., *Running enhances spatial pattern separation in mice.* Proc Natl Acad Sci USA, 2010. **107**(5): p. 2367-72.

Vaynman, S., Z. Ying, and F. Gomez-Pinilla, *Hippocampal BDNF mediates the efficacy of exercise on synaptic plasticity and cognition.* Eur J Neurosci, 2004. **20**(10): p. 2580-90.

Li, Y., et al., *TrkB regulates hippocampal neurogenesis and governs sensitivity to antidepressive treatment.* Neuron, 2008. **59**(3): p. 399-412.

199   **... aerobic exercise training increases the cells in other areas of the brain involved in major cognitive tasks:** Colcombe, S.J., et al., *Aerobic exercise training increases brain volume in aging humans.* J Gerontol A Biol Sci Med Sci, 2006. **61**(11): p. 1166-70.

Colcombe, S.J., et al., *Cardiovascular fitness, cortical plasticity, and aging.* Proc Natl Acad Sci U S A, 2004. **101**(9): p. 3316-21.

Rosano, C., et al., *Psychomotor speed and functional brain MRI 2 years after completing a physical activity treatment.* J Gerontol A Biol Sci Med Sci, 2010. **65**(6): p. 639-647.

Erickson, K.I., et al., *Physical activity predicts gray matter volume in late adulthood: the Cardiovascular Health Study*. Neurology, 2010. **75**(16): p. 1415-22.

Erickson, K.I., et al., *Aerobic fitness is associated with hippocampal volume in elderly humans*. Hippocampus, 2009. **19**(10): p. 1030-9.

Honea, R.A., et al., *Cardiorespiratory fitness and preserved medial temporal lobe volume in Alzheimer disease*. Alzheimer Dis Assoc Disord, 2009. **23**(3): p. 188-97.

Pereira, A.C., et al., *An in vivo correlate of exercise-induced neurogenesis in the adult dentate gyrus*. Proc Natl Acad Sci U S A, 2007. **104**(13): p. 5638-43.

Burdette, J.H., et al., *Using network science to evaluate exercise-associated brain changes in older adults*. Front Aging Neurosci, 2010. **2**: p. 23-23.

199 **Running in particular elevates cathepsin B:** Moon, H.Y., et al., *Running-Induced Systemic Cathepsin B Secretion Is Associated with Memory Function*. Cell Metab, 2016. **24**(2): p. 332-40.

200 **… they have very similar addictive physiological behaviour as morphine, heroine or nicotine:** Fernandes, R.M., et al., *The Effects of Moderate Physical Exercise on Adult Cognition: A Systematic Review*. Front Physiol, 2018. **9**: p. 667.

van den Berg, V., et al., *Physical Activity in the School Setting: Cognitive Performance Is Not Affected by Three Different Types of Acute Exercise*. Front Psychol, 2016. **7**: p. 723.

Best, J.R., et al., *Larger Lateral Prefrontal Cortex Volume Predicts Better Exercise Adherence Among Older Women: Evidence From Two Exercise Training Studies*. J Gerontol A Biol Sci Med Sci, 2017. **72**(6): p. 804-810.

Tsai, C.L., et al., *Impact of acute aerobic exercise and cardiorespiratory fitness on visuospatial attention performance and serum BDNF levels*. Psychoneuroendocrinology, 2014. **41**: p. 121-31.

Olson, R.L., et al., *Neurophysiological and behavioral correlates of cognitive control during low and moderate intensity exercise*. Neuroimage, 2016. **131**: p. 171-80.

200 **There is an emerging consensus that exercise in midlife prevents or delays dementia in later life:** Alty J, Farrow M, Lawler K. *Exercise and dementia prevention*. Pract Neurol, 2020 May;20(3): p. 234-240.

200 **As part of a research study, mice had genes modified so that they were more likely to get dementia:** Collins, A., et al., *Exercise improves cognitive responses to psychological stress through enhancement of epigenetic mechanisms and gene expression in the dentate gyrus*. PLoS One, 2009. **4**(1): e4330.

Choi, S.H., et al., *Combined adult neurogenesis and BDNF mimic exercise effects on cognition in an Alzheimer's mouse model*. Science, 2018. **361**(6406): eaan8821.

Maejima, H., et al., *Exercise and low-level GABA$_A$ receptor inhibition modulate locomotor activity and the expression of BDNF accompanied by changes in epigenetic regulation in the hippocampus.* Neurosci Lett, 2018. **685**: p. 18-23.

Moon, H.Y., et al., *Running-Induced Systemic Cathepsin B Secretion Is Associated with Memory Function*

201 **Background inflammation is closely linked to body fat … Regular physical activity decreases fat:** Ghilotti, F., et al., *Obesity and risk of infections: results from men and women in the Swedish National March Cohort.* Int J Epidemiol, 2019. **48**(6): p. 1783-1794.

Ross, R. and A.J. Bradshaw, *The future of obesity reduction: beyond weight loss.* Nat Rev Endocrinol, 2009. **5**(6): p. 319-25.

201 **Fat cells also make the immune response less efficient:** Lowder, T., D.A. Padgett, and J.A. Woods, *Moderate exercise protects mice from death due to influenza virus.* Brain Behav Immun, 2005. **19**(5): p. 377-80.

202 **… the need for mechanical ventilation in the ICU in patients with Covid-19 was seven times higher for those who were obese:** Simonnet, A., et al., *High Prevalence of Obesity in Severe Acute Respiratory Syndrome Coronavirus-2 (SARS-CoV-2) Requiring Invasive Mechanical Ventilation.* Obesity (Silver Spring), 2020. **28**(7): p. 1195-1199.

202 **… despite the evidence that exercise is a powerful tool to boost the immune system:** Sattar, N., I.B. McInnes, and J.J.V. McMurray, *Obesity Is a Risk Factor for Severe COVID-19 Infection.* Circulation, 2020. **142**(1): p. 4-6.

Centers for Disease Control and Prevention. *People of Any Age with Underlying Medical Conditions.* [2020 25 June 2020 17 July 2020]; Available from: https://www.cdc.gov/coronavirus/2019-ncov/need-extra-precautions/people-with-medical-conditions.html

202 **Exercising muscle releases enzymes called myokines that transiently block harmful inflammatory proteins:** Gulcelik, N.E., et al., *Adipocytokines and aging: adiponectin and leptin.* Minerva Endocrinol, 2013. **38**(2): p. 203-210.

Vieira-Potter, V.J., *Inflammation and macrophage modulation in adipose tissues.* Cell Microbiol, 2014. **16**(10): p. 1484-92.

Gleeson, M., et al., *The anti-inflammatory effects of exercise: mechanisms and implications for the prevention and treatment of disease.* Nat Rev Immunol, 2011. **11**(9): p. 607-15.

Ross, R. and A.J. Bradshaw, *The future of obesity reduction.*

202 **There is strong evidence to support the fact that it is never too late to start or to increase exercise:** Bartlett, D.B., et al., *Habitual physical activity is associated with the maintenance of neutrophil migratory dynamics in healthy older adults.* Brain Behav Immun, 2016. **56**: p. 12-20.

Timmerman, K.L., et al., *Exercise training-induced lowering of inflammatory (CD14+CD16+) monocytes: a role in the anti-inflammatory influence of exercise?* J Leukoc Biol, 2008. **84**(5): p. 1271-8.

Duggal, N.A., et al., *Major features of immunesenescence, including reduced thymic output, are ameliorated by high levels of physical activity in adulthood.* Aging Cell, 2018. **17**(2):e12750.

202 **Many studies of exercise performed one to six times per week over a period ranging from six weeks up to ten months have shown multiple positive effects:** Shimizu, K., et al., *Effect of moderate exercise training on T-helper cell subpopulations in elderly people.* Exerc Immunol Rev, 2008. **14**: p. 24-37.

Suchanek, O., et al., *Intensive physical activity increases peripheral blood dendritic cells.* Cell Immunol, 2010. **266**(1): p. 40-5.

Arner, P., et al., *Adipose lipid turnover and long-term changes in body weight.* Nat Med, 2019. **25**(9): p. 1385-1389.

203 **People over 65 are more prone not only to influenza but also to severe side effects from influenza:** Ciabattini, A., et al., *Vaccination in the elderly: The challenge of immune changes with aging.* Semin Immunol, 2018. **40**: p. 83-94.

203 **Unfortunately, the vaccine is not at all as effective at older ages as in young adults:** Osterholm, M.T., et al., *Efficacy and effectiveness of influenza vaccines: a systematic review and meta-analysis.* Lancet Infect Dis, 2012. **12**(1): p. 36-44.

Jefferson, T., et al., *Efficacy and effectiveness of influenza vaccines in elderly people: a systematic review.* Lancet, 2005. **366**(9492): p. 1165-74.

Siegrist, C.A. and R. Aspinall, *B-cell responses to vaccination at the extremes of age.* Nat Rev Immunol, 2009. **9**(3): p. 185-94.

203 **... doing aerobic exercise for the three months before having the flu vaccine significantly improved vaccination responsiveness:** Kohut, M.L., et al., *Moderate exercise improves antibody response to influenza immunization in older adults.* Vaccine, 2004. **22**(17-18): p. 2298-306.

Long, J.E., et al., *Vaccination response following aerobic exercise: can a brisk walk enhance antibody response to pneumococcal and influenza vaccinations?* Brain Behav Immun, 2012. **26**(4): p. 680-7.

203 **... regular physical activity is associated with these major health benefits:** Shepherd, S.O., et al., *Low-Volume High-Intensity Interval Training in a Gym Setting Improves Cardio-Metabolic and Psychological Health.* PLoS One, 2015. **10**(9): e0139056.

203 **... the majority of adults fail to meet the World Health Organization (WHO) recommended guidelines:** World Health Organization. *Global recommendations on physical activity for health.* [2010 May 6, 2020]; Available from: https://www.who.int/dietphysicalactivity/publications/9789241599979/en/

203 **... adults aged 40 and older reported spending more time on the toilet each week than walking**: UK Active. *Inactive Brits spend twice as long on toilet per week as they do exercising.* 2017 24 September 2017 May 7, 2020]; Available from: https://www.ukactive.com/events/inactive-brits-spend-twice-as-long-on-toilet-per-week-as-they-do-exercising/

205 **Protein supplementation in addition to muscle strengthening exercises is required:** Tessier, A.J. and S. Chevalier, *An Update on Protein, Leucine, Omega-3 Fatty Acids, and Vitamin D in the Prevention and Treatment of Sarcopenia and Functional Decline.* Nutrients, 2018. **10**(8):1099.

205 **... in addition to exercise, current recommendations are to stand where possible and to stand up at 45-minute intervals during prolonged sitting:** Miller, K.J., et al., *Comparative effectiveness of three exercise types to treat clinical depression in older adults: A systematic review and network meta-analysis of randomised controlled trials.* Ageing Res Rev, 2020. **58**: 100999.

Harris, T., et al., *Effect of pedometer-based walking interventions on long-term health outcomes: Prospective 4-year follow-up of two randomised controlled trials using routine primary care data.* PLoS Med., 2019. **16**: e1002836.

GreyMatters. *Stand Up For Your Brain.* [2019 13 May 2021]; Available from: https://greymattersjournal.com/stand-up-for-your-brain/

Jung, J.-Y., H.-Y. Cho, and C.-K. Kang, *Brain activity during a working memory task in different postures: an EEG study.* Ergonomics, 2020. **63**(11): p. 1359-1370.

205 **This is good for 'waking up' our physiological systems**: Maasakkers, C., Kenny R.A., et al., *Hemodynamic and structural brain measures in high and low sedentary older adults.* J. Cereb. Blood Flow Metab. 2021 Oct;41(10):2607-2616

205 **... a combination of aerobic and muscle strengthening exercises coupled with regular standing-up when sitting for prolonged periods are preferred:** Davidsen, P.K., et al., *High responders to resistance exercise training demonstrate differential regulation of skeletal muscle microRNA expression.* J Appl Physiol (1985), 2011. **110**(2): p. 309-17.

205 **... sarcopenia is relatively new in medicine ... It is a progressive and generalised ageing muscle disease:** Marzetti, E., et al., *Sarcopenia: an overview.* Aging Clin Exp Res, 2017. **29**(1): p. 11-17.

Cruz-Jentoft, A.J., et al., *Sarcopenia: revised European consensus on definition and diagnosis.* Age Ageing, 2019. **48**(1): p. 16-31.

Vellas, B., et al., *Implications of ICD-10 for Sarcopenia Clinical Practice and Clinical Trials: Report by the International Conference on Frailty and Sarcopenia Research Task Force.* J Frailty Aging, 2018. **7**(1): p. 2-9.

206　**We lose 15 per cent of muscle strength due to decline in muscle mass every ten years:** McLean, R.R. and D.P. Kiel, *Developing Consensus Criteria for Sarcopenia: An Update.* J Bone Miner Res, 2015. **30**(4): p. 588-592.

Limpawattana, P., P. Kotruchin, and C. Pongchaiyakul, *Sarcopenia in Asia.* Osteoporosis Sarcopenia, 2015. **1**.

206　**Studies vary regarding how common sarcopenia is but some estimates are that up to two thirds of people over 70 have it:** Nascimento, C.M., et al., *Sarcopenia, frailty and their prevention by exercise.* Free Radic Biol Med, 2019. **132**: p. 42-49.

Siparsky, P.N., D.T. Kirkendall, and W.E. Garrett, Jr., *Muscle changes in aging: understanding sarcopenia.* Sports Health, 2014. **6**(1): p. 36-40.

206　**... if you have a bad cold and are confined to bed for a few days, be aware to put effort into keeping muscles moving**: Morley, J.E., *Frailty and Sarcopenia: The New Geriatric Giants.* Rev Invest Clin, 2016. **68**(2): p. 59-67.

Frederiksen, H., et al., *Hand grip strength: a phenotype suitable for identifying genetic variants affecting mid- and late-life physical functioning.* Genet Epidemiol, 2002. **23**(2): p. 110-22.

Marzetti, E., et al., *Sarcopenia: an overview.*

Nascimento, C.M., et al., *Sarcopenia, frailty and their prevention by exercise*

Kalinkovich, A. and G. Livshits, *Sarcopenic obesity or obese sarcopenia: A cross talk between age-associated adipose tissue and skeletal muscle inflammation as a main mechanism of the pathogenesis.* Ageing Res Rev, 2017. **35**: p. 200-221.

206　**Whereas aerobic exercise is a must, it is not sufficient without additional resistance exercises:** Fragala, M.S., et al., *Resistance Training for Older Adults: Position Statement From the National Strength and Conditioning Association.* J Strength Cond Res, 2019. **33**(8): p. 2019-2052.

206　**... loss of muscle mass is generally gradual**: Melton, L.J., 3rd, et al., *Epidemiology of sarcopenia.* J Am Geriatr Soc, 2000. **48**(6): p. 625-30.

206　**... those who have been involved in physical activity from an early age do have an advantage:** Gallagher, D., et al., *Appendicular skeletal muscle mass: effects of age, gender, and ethnicity.* J Appl Physiol (1985), 1997. **83**(1): p. 229-39.

Janssen, I., et al., *Skeletal muscle mass and distribution in 468 men and women aged 18-88 yr.* J Appl Physiol (1985), 2000. **89**(1): p. 81-8.

Frontera, W.R., et al., *Aging of skeletal muscle: a 12-yr longitudinal study.* J Appl Physiol (1985), 2000. **88**(4): p. 1321-6.

Goodpaster, B.H., et al., *The loss of skeletal muscle strength, mass, and quality in older adults: the health, aging and body composition study.* J Gerontol A Biol Sci Med Sci, 2006. **61**(10): p. 1059-64.

206 **Resistance exercise mitigates the effects of ageing on the nerves that feed skeletal muscles:** Fragala, M.S., et al., *Resistance Training for Older Adults.*

206 **At a cellular level, oxidative stress is improved:** Johnston, A.P., M. De Lisio, and G. Parise, *Resistance training, sarcopenia, and the mitochondrial theory of aging.* Appl Physiol Nutr Metab, 2008. **33**(1): p. 191-9.

207 **A programme should include an individualised, periodised approach … start up again as soon as you can:** McGrath, R.P., et al., *Muscle Strength Is Protective Against Osteoporosis in an Ethnically Diverse Sample of Adults.* J Strength Cond Res, 2017. **31**(9): p. 2586-2589.

McLean, R.R., et al., *Criteria for clinically relevant weakness and low lean mass and their longitudinal association with incident mobility impairment and mortality: the foundation for the National Institutes of Health (FNIH) sarcopenia project.* J Gerontol A Biol Sci Med Sci, 2014. **69**(5): p. 576-583.

Peterson, M.D., et al., *Muscle Weakness Thresholds for Prediction of Diabetes in Adults.* Sports Med, 2016. **46**(5): p. 619-28.

Dalsky, G.P., et al., *Weight-bearing exercise training and lumbar bone mineral content in postmenopausal women.* Ann Intern Med, 1988. **108**(6): p. 824-8.

Nelson, M.E., et al., *Effects of high-intensity strength training on multiple risk factors for osteoporotic fractures. A randomized controlled trial.* JAMA, 1994. **272**(24): p. 1909-14.

Westcott, W.L., *Resistance training is medicine: effects of strength training on health.* Curr Sports Med Rep, 2012. **11**(4): p. 209-16.

Shaw, C.S., J. Clark, and A.J. Wagenmakers, *The effect of exercise and nutrition on intramuscular fat metabolism and insulin sensitivity.* Annu Rev Nutr, 2010. **30**: p. 13-34.

Bweir, S., et al., *Resistance exercise training lowers HbA1c more than aerobic training in adults with type 2 diabetes.* Diabetol Metab Syndr, 2009. **1**: 27.

207 **… only 8 per cent of adults over 75 years of age in the United States participate in muscle-strengthening, resistance activities:** National Center for Health Statistics (NCHS), *National Health Interview Survey, 2015.* 2016, Centers for Disease Control and Prevention (CDC): Hyattsville, Maryland.

207 **Reported barriers to participation include fear, health concerns, pain, fatigue, lack of social support:** Burton, E., et al., *Motivators and Barriers for Older People Participating in Resistance Training: A Systematic Review.* J Aging Phys Act, 2017. **25**(2): p. 311-324.

207 **If you do not at present do resistance exercises to complement aerobic exercise then I recommend that you start:** Bunout, B., et al., *Effects of nutritional supplementation and resistance training on muscle strength in free living elders. Results of one year follow.* J Nutr Health Aging, 2004. **8**(2): p. 68-75.

Pahor, M., et al., *Effects of a physical activity intervention on measures of physical performance: Results of the lifestyle interventions and independence for Elders Pilot (LIFE-P) study.* J Gerontol A Biol Sci Med Sci, 2006. **61**(11): p. 1157-65.

Latham, N.K., et al., *Effect of a home-based exercise program on functional recovery following rehabilitation after hip fracture: a randomized clinical trial.* JAMA, 2014. **311**(7): p. 700-8.

207   ... **even in people 90 and over, resistance exercises are feasible and make a difference:** Papa, E.V., X. Dong, and M. Hassan, *Resistance training for activity limitations in older adults with skeletal muscle function deficits: a systematic review.* Clin Interv Aging, 2017. **12**: p. 955-961.

208   ... **protein supplements should be used to complement resistant exercise programmes:** Kimball, S.R. and L.S. Jefferson, *Control of protein synthesis by amino acid availability.* Curr Opin Clin Nutr Metab Care, 2002. **5**(1): p. 63-7.

Dardevet, D., et al., *Stimulation of in vitro rat muscle protein synthesis by leucine decreases with age.* J Nutr, 2000. **130**(11): p. 2630-5.

Hasten, D.L., et al., *Resistance exercise acutely increases MHC and mixed muscle protein synthesis rates in 78-84 and 23-32 yr olds.* Am J Physiol Endocrinol Metab, 2000. **278**(4): p. E620-6.

Balagopal, P., et al., *Effects of aging on in vivo synthesis of skeletal muscle myosin heavy-chain and sarcoplasmic protein in humans.* Am J Physiol, 1997. **273**(4): p. E790-800.

208   **the group who were treated with daily whey protein which was leucine (an amino acid) and vitamin D enriched for a three month period had a significant improvement:** Robinson, S., C. Cooper, and A. Aihie Sayer, *Nutrition and Sarcopenia: A Review of the Evidence and Implications for Preventive Strategies.* J Aging Res, 2012. **2012**: 510801.

Tessier, A.J. and S. Chevalier, *An Update on Protein, Leucine, Omega-3 Fatty Acids, and Vitamin D in the Prevention and Treatment of Sarcopenia and Functional Decline.*

208   **Animal and human experimental studies show that vitamin E benefits new muscle formation:** Chung, E., et al., *Potential roles of vitamin E in age-related changes in skeletal muscle health.* Nutr Res, 2018. **49**: p. 23-36.

Acknowledgements

I would like to thank the following people:

My husband, Gary, for his patience and insight and my sons, Redmond and Pearse Traynor, who both helped with editing (thanks to Pearse for his persistent deep dive into the facts and his thorough feedback from a twentysomething standpoint). My sisters Kate, Paula and Grace with whom I shared lots of laughs and tears during the writing process.

The fabulous TILDA team with whom I have worked so closely over the past 15 years, including Dr Silvin Knight for some graphics, Dr Cathal McCrory for modifyingTILDA tests, Deirdre O'Connor and Eleanor Gaffney for administrative support for the book.

The TILDA participants who generously give so much of their time to enable a great study which has contributed to our understanding of the process of ageing and helped to change policy and practice globally.

Daniel McCaughey who ably assisted with literature searches

and data reviews whilst studying medicine; and to my caring mentors, Professor Richard Sutton and Professor Davis Coakley.

My secretary for the past 15 years, Helen Fitzpatrick, for her patience, wisdom and hard work assisting me with *Age Proof*.

I'd also like to thank Bill Hamilton, my literary agent, and the great team at Bonnier with whom I have very much enjoyed working.

I have had a very privileged life in medicine and continue to love every minute of it. Thank you to all of the patients who have shared their lives and experiences in order to broaden mine.

PICTURE CREDITS

Page 15
Bar chart reproduced from:
Belsky, D.W., Caspi, A., Houts, R., Cohen, H.J., Corcoran, D.L., Danese, A., Harrington, H., Israel, S., Levine, M.E., Schaefer, J.D. and Sugden, K, *Quantification of biological aging in young adults*. PNAS 112(30); issued July 28th 2015 page 4105, Figure 2

Pages 29 and 30
Photographs of nuns reproduced courtesy of the School Sisters of Notre Dame North American Archives, Milwaukee, Wisconsin

Page 47
Image of chromosome reproduced courtesy of 123rf.com

Page 99
Image of human head and brain reproduced courtesy of 123rf.com

Page 133
Image of cell reproduced courtesy of 123rf.com

Page 161
Photographs of ageing monkeys reproduced by permission of the American Association for the Advancement of Science, from the publication *Science*

Index